FIXED ON NITROGEN

A SCIENTIST'S SHORT STORY

DAVID R DENT

ADG
PUBLISHING

ISBN: 978-1-9162181-0-9 (Paperback edition - b&w)
ISBN: 978-1-9162181-1-6 (Paperback edition - colour)
ISBN: 978-1-9162181-2-3 (Kindle eBook edition)

Front cover and book design by Weavebean Creative

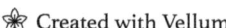

Created with Vellum

CONTENTS

"A man who dares to waste one hour of time has not discovered the value of life"

Charles Darwin

ACKNOWLEDGMENTS

The story in this book would not have been possible without the dedication, hard work and scientific abilities of the scientists and technical staff of Azotic Technologies Ltd past and present. My heartfelt thanks and genuine appreciation goes to each and every one of them: Gary Devine, Nathalie Narraidoo, Nick Gosman, Inma del Castillo, Mike Thomas, Katrin Schwarz, Pedro Carvalho, Emma Smail, Michelle Nuttall, Erika Wagner, Thomas Reed, Mariya Iqbal, Charles Owen, Ian Clarke, Zoe Dunsiger, Dhaval Patel and Thomas MacCalman plus the numerous interns and associates for all of their contributions. Of particular note I want to thank those individuals who took-on board extra leadership responsibilities and duties and committed to delivering with real entrepreneurial spirit, and who made a real difference: Gary Devine (Scientist and Laboratory Manager), Pedro Carvalho (Chief Crop Scientist), Erika Wagner (USA Chief Agronomist) and Nathalie Narraidoo (Scientist). Thanks also Alan Burbidge for comments on an early draft and to my daughter Catherine for all her copying editing and for removing so many exclamation remarks!

Dedicated to
Professor Edward C. Cocking
"A Truly Extraordinary Scientist"

PREFACE

I have written any number of books; text-books of various kinds and lengths, a novel, and books of poetry, but of all of them, I have never questioned quite as much as I have with this memoir. The reason for my question I suspect is down to the simple fact that although some of this is biographical, a good proportion of it is autobiographical - it is certainly not fictional. In such a book there is no hiding behind a scientist's dry facts, theory and practice, it inevitably gives some insight into who I am, a rather private individual who is forgoing some of that privacy, which scares the living daylights out of me. That said, for the most part, I have written this book, because I believe it is a story that needs to be told, it represents the sort of narrative that rarely gets written of science and innovation and some of the people who make extraordinary things happen.

While it is certainly important to record the facts, it is also important to learn about the process, of how the decisions were made that led to the generation of outcomes, especially where some of them are unanticipated. This is not a scientific story all about a neatly controlled systematic stepwise linear process where x outcome led to experiment y where z was anticipated and z was delivered. Life is not like that, scientific research is certainly not like that, and innovation rarely so. Rather it is a bewildering process of trial and of human error, of misdirection and miscalculations, of taking risks based on incomplete information, of being prepared to put ones 'head on the block' and living with any consequence that may arise because of this, of the ridiculous time constraints and the unreasonable and constant pressure to deliver and the sleepless nights this generates, day after day, week after week and month after month; of being scared, being lonely and overworked. Such are the realities that underpin the story that is told in this book. Of course these challenges are balanced by the moments of sheer fun and delight associated with an unexpected outcome and a benefit beyond our wildest imaginations. It is not appropriate as a scientist to talk about luck but given what I have declared above you can see why it might seem appropriate sometimes. And we certainly had our measure of luck - or serendipity as we perhaps more euphemistically refer to it in scientific circles.

This book also provides an opportunity to put on record and recognise the achievements, to celebrate and to thank the team of scientists who made this story what it is, as well as to pay a real and lasting tribute to the pioneering and paradigm shattering insights and research of Prof Ted Cocking. Insights that only now being recognised for what they are – insights that are scientifically important but also addressing an aspect of life that is crucial to the future of our planet.

Nitrogen fertilisers and their replacement with more sustainable alternatives have been a long running scientific problem for scientists against which a number of potential solutions exist. Sadly were it only a scientific problem, it would not perhaps matter a great deal, but the world has gone beyond its nitrogen boundary and a large part of the problem is the greenhouse gas (GHG) nitrous oxide emissions from agriculture. The climate change clock is not only ticking - the alarm has now gone off - and we have been caught over sleeping! Now is not the time to prevaricate and walk away from risk, from something different and contrary to established thought. Now is the time to embrace disruptive climate smart technologies. The development of the nitrogen-fixing technology described in this story provides one of the solutions to this global nitrogen problem. Yet the struggle to gain any traction despite its obvious promise is an abject lesson of the intractability of scientists, philanthropists, governments and business towards genuine innovation and a testament to their attitudes towards risk. One thing for certain is that the technology related here represents part of a solution to our global nitrogen problem. I hope that this story puts on record the means by which the journey began and, how when the opportunity became evident, a number of individuals embraced its potential and the promise it offered.

Finally I would like to acknowledge one or two of life's simple maxims, firstly 'one can achieve anything if you are prepared to make the sacrifice to make it possible' and secondly, that 'our world really can change in extraordinary ways for the greater good through the hard work, courage, tenacity and dynamism of just a few individuals'. This book is a tribute to my scientific colleagues at Azotic Technologies Ltd who over the period this book covers displayed these qualities as individuals and as a team. We fought against the odds, got it right just often enough and I salute them for the significance and scale of their achievement.

David Dent PhD, FRSB

FOREWORD

The conundrum of reactive nitrogen such as more nitrogen fertiliser use to feed 9.7 billion people by 2050 on one hand, and reducing fertiliser use to lower green house gas (GHG) emissions and nitrate pollution on the other, poses a real threat to our food security. It is estimated that around 40 per cent of global population gets sustained food due to 500 million tones of nitrogenous fertilisers used in agriculture each year. At the same time, the nitrogen fertiliser production technology and application methods are quite inefficient and have major implications on soil health and environmental sustainability. It is estimated that production of nitrogenous fertilisers burns almost 3 per cent of world's natural gas and releases around 3 per cent of total carbon emissions. Globally, 50-70 per cent of all nitrogenous fertilisers is wasted or lost, either to the atmosphere as harmful greenhouse gas as nitrous oxide, or into waterbodies/groundwater as nitrate. Thus, in general, reactive nitrogen (ammonia, nitrate, nitrite and nitrous oxide) have a cascading effect on soil, environment, livestock and human health. The sheer magnitude and complexity of the reactive nitrogen problem makes solutions quite difficult and, therefore, need science-based disruptive innovation as an effective alternative.

Nitrogen-fixing bacteria, that normally are found in leguminous crop plants, do utilise non-reactive N2 and convert it into the form that can be used by plants, thus reducing the dependence on nitrogenous fertilisers. These bacteria form root nodules especially in legumes like peas, lentil, chickpea, soybean etc. Hence, if a species of bacteria could be found in nature which could fix nitrogen in the cereal crops like wheat, rice, maize etc. then it would revolutionise global agriculture with less dependence on costly chemical fertilisers. It is in this context, biological nitrogen fixation (BNF) has received considerable attention as a flagship programme by the global scientific community for more than five decades since the Green Revolution period but unfortunately with no real breakthrough.

In this context, the last four decades of continuous research by Professor Edward (Ted) Cocking, FRS had been all about finding a solution to this major challenge. From studies trying to identify ways of breaking through the hard outer wall of plant cells to allow investigation of inner cell workings, Ted had found a way to identify bacteria that had the characteristics to enter plant cells naturally. If he could then find a bacteria with not only the ability to break down plant cell walls, but to also fix nitrogen, it would have been quite a unique story - truly transformative and rather revolutionary.

In this book entitled ***Fixed on Nitrogen***, David Dent highlights two contrasting approaches taken by himself and Professor Ted to solve the nitrogen fertiliser problem through non-reactive nitrogen that occupies about 80 percent of our atmosphere. As already stated, this form of nitrogen does not negatively impact our environment and thus would offer hope for a sustainable solution to global nitrogen problem. David Dent relates the story of just how circumstances conspired to make such a thing possible. Ted and David approaching life and their career from two completely different perspectives; Ted a well distinguished university professor and David being a scientist cum dynamic entrepreneur came together to develop something that is much bigger than either could have created alone. Ted's academic expertise is able to find expression within the business context and product orientated approach that David exemplifies. David's, is an experience and expertise that has not been honed in the singular focused pursuit within an academic discipline or institute but rather quite the opposite - a varied career as an inventor, post-doctoral researcher, an SME product development manager, an international research programme manager, a Director of an inter-governmental organisation, a consultant, a government advisor and an entrepreneur. Both of them have been recognised by the Royal Society, Ted as a Fellow and David as one of their Entrepreneurs in Residence.

Indeed it is highly satisfying to see that their joint efforts have culminated in such a disruptive innovation, which is being scaled now to help small-holder farmers around the world to increase their production without much dependence on costly nitrogenous fertilisers. Both Professor Edward Cocking and Dr David Dent deserve great appreciation for this breakthrough of such global significance.

R. S. Paroda
Founder Chairman, Trust for Advancement of Agricultural Sciences (TAAS); Former Director General, Indian Council of Agricultural Research (ICAR); Secretary, Department of Agricultural Research and Education (DARE), Government of India.

1

WHERE THERE IS A BEGINNING ...

"Once upon a time" ... in Bangkok

There are not many stories about science and innovation that begin in a bar in Bangkok, Thailand, but this, I cannot lie, is one of them. The bar in question was a hotel bar, located on a lower floor of the Grand Hyatt Erawan Hotel in Ratchaprasong, a luxurious hotel within the very heart of the great bustle of central Bangkok. The reason for my presence in the bar at all was as a result of an event earlier that day in Grand Ballroom 3, where I had been an organiser and speaker at a workshop entitled *Innovation, Industry and Investment:* a workshop sponsored by the UK Government Department for Trade and Investment (UKTI) in Thailand. It was the 8th March 2011 and the person who sat opposite me was Dr Alan Burbidge of the University of Nottingham, also a speaker at the workshop. This event was the first time Alan and I had ever met and the conversation we were having over a beer was happening because of a topic I had mentioned in my presentation that had struck a chord with Alan in his role as the University Licensing Executive. Little did we understand at that time the scale of the repercussions that would arise from that simple enquiry and those few drinks on that night in a bar in Bangkok!

The topic of our conversation at this moment may seem for you, neither inspiring nor significant. Even under the influence of any numbers of beers most people in the world, would probably hold a similar view. But to me then, as now I was intrigued and beguiled with a vision of the possibilities that Alan's story appeared to offer. Even as I say this now, it still seems rather tame. This was not about the scientific glamour of sub-atomic particles, the unfathomable scale of the cosmos or the mesmerising intricacies of evolutionary theory, but rather concerned a subject referred to somewhat simply as, biological nitrogen fixation or BNF, for short.

For those not familiar with the subject, and for those who wish to continue past these first few paragraphs; biological nitrogen fixation (BNF), is a natural process by which some bacteria are able to take some of the 78% of nitrogen that makes up our air and turn it into ammonia, a form of nitrogen that can be used by plants to help them grow. Nitrogen is usually difficult for plants to obtain and without it they cannot produce the enzymes, proteins and DNA that they need to grow and to produce our food. If a way can be found to harness biological nitrogen fixation through these nitrogen-fixing bacteria then we could improve crop plant yields and potentially replace the use of mineral nitrogen fertilisers.

There are a number of different types of bacteria that are able to fix nitrogen, and many of them exist in the soil and are known as free-living nitrogen-fixers. Another category of bacteria (although they can also live freely in the soil), form close associations with plants which can be very specific to particular plant species. The most well known of these bacteria are called Rhizobia and they have such close associations with their hosts that the plants form special root structures called 'nodules' in order to house the bacteria as part of their roots. These structures provide a safe home for the bacteria for them to fix nitrogen obtained from the air pockets in the soil and pass on some of this to the plant. Thus legumes are known as mostly self-sufficient in nitrogen and do not require the addition of mineral nitrogen fertiliser for them to thrive. Unfortunately these specialist associations with Rhizobia are restricted to the leguminous plants, the peas and the beans, including soybean; nitrogen fertilisers do not have to be added to pea and bean crops for them to produce high yields.

A third group of nitrogen-fixing bacteria are those that are not free-living in the soil; they need a plant to inhabit in order to survive, and unlike Rhizobia they do not form specialised root structures with the plants they colonise. These types of bacteria are known as obligate endophytes (obligate meaning 'compelled' and endophyte meaning 'living within the plant') and they have generally been less well studied than the Rhizobia.

The reason Alan and I were in discussion was the praise I had proffered during my presentation of the fundamental research undertaken by the scientists working on Rhizobia and nitrogen fixation at the UK Research Council supported John Innes Centre (JIC) in Norwich UK. Research at the JIC had been attempting for a number of years to understand the mechanisms involved in root nodule formation with Rhizobia with the idea of transferring the capability from peas and beans to cereal crops, particularly wheat. It was hoped that JIC's molecular and genetic manipulation approach could in the future produce wheat plants with root nodules and rhizobia to fix nitrogen, thereby reducing the need for nitrogen fertilisers for global cereal production. This was not the only research in this area, but having previously been involved in its promotion in Thailand I had naturally used it as an example in my presentation.

At the end of my presentation, Alan had approached me and said, "if you think the work at the John Innes Centre is exciting then you need to hear about the research of Ted Cocking at Nottingham University", after which he nonchalantly sauntered off to talk with other colleagues. I was aware of Professor Cocking's work and indeed had highlighted its potential in a report I had written recently but had clearly, if Alan's words were to be believed, not fully appreciated its significance. My curiosity had been sparked, and I needed to know more, and I arranged to meet Alan later that evening in the bar to find out what I obviously had not picked up about Professor Cocking's research.

So what - it's only nitrogen?

It is perhaps worth relating at this point why this nitrogen fixation phenomenon matters at all, and why research effort is being expended to ensure biological nitrogen fixation can be extended from peas and beans to cereal crops.

Firstly, every plant requires nitrogen to be able to grow, and under natural conditions limitations on the availability of nitrogen can stunt or constrain a plant's growth. In order to overcome this limitation in agriculture we apply different types of nitrogen. In our past, and in organic systems, we have utilised manure from farm animals but since the invention of the Haber-Bosch process by Fritz Haber and Carl Bosch in 1910 we have been able to produce nitrogen in the form of ammonia that can be used to make nitrate and ammonium based nitrogen fertilisers.

The present production capacity for synthetic ammonia utilising technologically enhanced versions of the Haber-Bosch process is over 175 million metric tonnes per year and growing at a rate of 1-2% a year (Brightling, 2018). Approximately 85% of this tonnage is used as fertiliser for food production. What seemed like an ingenious solution to the need for fertiliser a hundred years ago is now running into serious problems. This is because, the Haber-Bosch process burns 3% of the world's production of natural gas and releases approximately 3% of the world's carbon emissions (Zhang 2016) . If relying on fossil fuels to give the world electricity and heat is unsustainable, so is relying on fossil fuels to grow our food. Any reduction in these gases through use of an alternative to ammonia based fertilisers would have a significant global impact reducing the threat of climate change.

Currently, over 500 million tonnes of ammonia are produced each year through the Haber-Bosch process in order to meet the global needs for nitrogen fertiliser. This tonnage accounts for roughly 1% of the world's energy usage and 3-5% of natural gas usage (Smith, 2002). This would be fine if there were no negative side effects to the production and use of ammonia in this way. Unfortunately, some of these problems are now considered so great that the value of nitrogen fertilisers to society has come into question.

In addition to this a key point to understand is that on average, for every 100 units of nitrogen used in global agriculture, only 17 are consumed by humans as crop, dairy or meat products (UNEP, 2007). Global nitrogen-use efficiency of crops, as measured by recovery efficiency in the first year (fertilised crop nitrogen uptake minus unfertilised crop nitrogen uptake/nitrogen applied), is generally less than 50% under most on-farm conditions (Tilman *et al.*, 2002; Doberman, 2007) and can be as low as 30%. This means that between 50 and 70% of all fertiliser is wasted or lost, either to the atmosphere as a harmful greenhouse gas - nitrous oxide, or into waterways as nitrate run-off. Nitrogen fertiliser application for the purposes of feeding our crops is a very inefficient and wasteful process that has major implications for climate change and sustainability of our global food production systems.

Firstly, nitrous oxide is 300 times more potent a greenhouse gas (GHG) than carbon dioxide and agriculture represents its largest source (Reay *et al.*, 2012). Nitrous oxide (N_2O) traps more heat over all time frames compared with methane on a weight basis [100 year global warming potential (GWP100) of 298 vs. 34; GWP20 of 268 vs. 86] and has a longer atmospheric lifetime (121 vs. 12 years; Myhre *et al.*, 2013). All in all, nitrous oxide is a major GHG problem and fertiliser use in the UK for example, contributes two thirds of the total climate changes gases from agriculture (CCC, 2017). Nitrate pollutes waterways and causes excess nutrients in watersheds, destroying wildlife, ecosystems, businesses and livelihoods dependent upon these. Nitrate run-off from agricultural lands also contaminates drinking water, with more than 1.5 million Americans drinking water from wells contaminated with nitrate and 5% of the European population exposed to unsafe levels (van Grinsven *et al.*, 2006). Addressing the pollution problems and the damage caused by nitrogen fertiliser use is currently expensive, making it unsustainable. Removing nitrates from water in Europe has been estimated to cost each taxpayer between £130-650 per year. The economics of fertiliser use in European agriculture is questionable, given that the overall nitrous oxide damage costs £60-80 billion per year, a sum more than double the extra income gained from using nitrogen fertilisers in agriculture in Europe (Sutton *et al.*, 2011).

The projected increased demand for food as we try to feed the burgeoning world population will require an overall increase in fertiliser nitrogen requirements from (for example) around 100 Mt to 135 Mt N by 2030 (Erisman *et al.*, 2008), as well as other mitigation measures. Large improvements in nitrogen-use efficiency are certainly possible especially given the excessive use of fertiliser in some countries, and other measures such as changing the source of N, using fertilisers stabilised with urease or nitrification inhibitors or slow- or controlled-release fertilisers, and through optimising nitrogen fertiliser placement and timing (Flynn & Smith, 2010; Synder *et al.*, 2009; Del Grosso & Grant, 2011). Reducing the amount of nitrogen fertiliser applied by using biological nitrogen fixation could make a significant contribution to the nitrous oxide emissions problem globally.

On the upside however, the fertilisers produced by the Haber Bosch process have made it possible to feed a growing world population for the last Century. Certainly there would have been famine and deprivation on a massive scale without nitrogen fertiliser use to produce food as the world's population grew from 1.6 billion in 1900 to 6 billion by 2000. It has also been claimed that the current use of 500 million tonnes of nitrogen fertiliser produced each year manages to sustain around 40% of the global population (Smil, 1999; Fryzuk, 2004). With the world's population set to grow to a staggering 9.7 billion by 2050, removing the use of fertilisers at this time would seem to represent a genuine threat to world food security.

We therefore appear to face a conundrum: on the one hand we require ever more nitrogen fertiliser in order produce the high crop yields to feed the world's growing population, while at the same time we need to reduce fertiliser use in order to lower GHG emissions and nitrate pollution of our waterways. While other mitigation methods will contribute to reducing the nitrous oxide emissions from agriculture, a more sustainable source of nitrogen is required. Biological nitrogen fixation, has the potential to become such a source of nitrogen, enabling and meeting the twin global needs of both climate smart agriculture and food security.

Why hasn't it been done already?

Biological nitrogen fixation with Rhizobia has been known and used in agriculture for over a century, through the production and addition of Rhizobia based inoculants to pea and bean crops. However, transferring this capability to modern intensive agriculture even for soybean, a major food crop, has proved challenging, while scientists have set their sights on being able to transfer the nitrogen fixation of Rhizobia into the major cereal crops.

This idea of creating a cereal crop, such as maize, wheat, or rice, that could fix nitrogen first gained credence and some momentum in the late 1960s. It was an alluring idea for both scientists and policy-makers concerned with issues of agricultural production, especially in developing countries, and the environmental and economic problems associated with the increasing use of nitrogen fertiliser. Circumstances conspired in favour of research funding for nitrogen fixation when in the early 1970's fuel prices increased which led to increases in the cost of fertilisers raising concerns about food security. A range of people including scientists from the BNF community envisaged a new opportunity and articulated a vision whereby BNF for cereals could herald a new era of sustainable agriculture. In their enthusiasm they collectively enunciated a promise; a promise that BNF and nodulation research would lead to the creation of a nitrogen-fixing cereal *in the near future* (Simmonds, 2008). So persuasive and appealing was this promise that funding from all manner of sources was subsequently directed at BNF and nodulation research, with a specific emphasis on BNF molecular genetics and plant molecular biology - new techniques at the forefront of science at the time. Hence for many years, optimism and expectations were running high with the potential opportunity offered by BNF and the benefits that could accrue from its widespread application.

Winding forward 40 years and with the benefit of hindsight, this promise based on a molecular approach to generating root nodules on cereals has yet to be realised and it is now generally thought that the promise of the 70's for molecular based BNF research was very much overstated (Simmonds, 2008; Rogers & Oldroyd, 2014). However, with the advances that have been made over this period, and these have been substantial, "... we are now [finally] in a good position to initiate this engineering approach" (Rogers & Oldroyd, 2014). However, the reality is that such root nodule-based and Rhizobia-associated nitrogen fixation in cereals is probably another 10-15 years away. Science sometimes can take time to deliver practical solutions to particular problems.

Pesky Paradigms

The problem of establishing root nodules and nitrogen fixation in cereals has proved a difficult matter to address over the years using a molecular based approach, but science is all about solving puzzles and finding solutions to intractable problems. However, the scientific process is not always straightforward or a simple sequential process based on obvious, small incremental steps. On occasion, progress may need to occur in bigger, more significant jumps and sometimes something novel or 'off the wall' is required to generate significant new developments and changes of approach.

The physicist Thomas Kuhn first defined the process of scientific progress in a book called "The Structure of Scientific Revolutions" (Kuhn, 1962). Kuhn argued that 'normal science' is the incremental process whereby scientists work as a community with a common way of thinking, approach, goal or established scientific opinion that is shared by them all - a structure Kuhn called a 'paradigm' (Dent, 1995). In this phase of scientific process most researchers are interested in solving some of the anomalies that occur within the paradigm - their main driver is maintaining the *status quo* and uncovering what they expect to uncover. This is because the 'paradigm' represents a limited framework within which each specialist subject is explored. A framework limited by (i) creating problems that do not exist by adherence to particular divisions, polarisations, conceptualisations; (ii) acting as conceptual traps or prisons which prevent a more useful re-arrangement of information; and (iii) through blocking by adequacy - being just sufficient to address a problem (de Bono, 1970). Eventually when too many unresolved anomalies occur and the limitations and constraints of a paradigm become overwhelmingly 'apparent' and cannot be ignored, then a 'paradigm shift' may occur when an extraordinary scientist or scientific phenomenon supplants the prevailing paradigm with a new orthodoxy (Medawar, 1981).

It is easy to see how BNF and the nodulation research community and their molecular-based approach over the last 40 years represents a paradigm; one in which a number of assumptions have been made about the type of bacteria required, the nature of interactions between the plant and the bacteria, the requirement for root nodules and the conditions needed for nitrogen fixation. The question has to arise, given the length of time taken and the time still required to deliver a practical working solution for BNF in cereals, whether a paradigm shift is required - or indeed whether there is one already in the making?

Meanwhile Back in Bangkok...

Anyone who has attended a conference will know that often most of the important developments, decisions and contacts are made and relationships built during the social interactions in and around the main event. So from that point of view it was nothing unusual for Alan and I to get together in the bar to start a discussion that would influence our lives in the short term and the potential to transform the lives of others in the future. I have always quite liked the local Thai Singha beer and with a chilled glass in front of both Alan and I, we talked about nitrogen-fixation. Alan explained how Edward Cocking, or Ted as he is known to most, had been working with a species of bacteria called *Gluconacetobacter diazotrophicus* which had been isolated from sugarcane where it had been shown to fix nitrogen. Going beyond this, Ted's research had demonstrated that under the right conditions this bacterium could be encouraged to colonise roots of a range of crop plant species, including the cereals wheat and maize, and to move throughout the plant fixing nitrogen. However, the key point that had led the University to patent the methodology was that the bacteria colonised the plants intracellularly - meaning they actually entered inside the plant cells where they formed a stable relationship with the plant - known as a symbiotic interaction (where both partners benefit from the association). What was exciting about this claim is that while at this time bacteria were known to colonise animal cells intracellularly, it was not accepted that such intracellular colonisation could generally occur in plants! Making this claim and the practical demonstration of this in the patent application was scientifically revolutionary - a first - and incredibly exciting. The claims made in the University patent that the bacteria once they were intracellular could then fix nitrogen, whether they were in the roots or leaves of the colonised plant, was just incredible! Alan and I had a long and fruitful discussion in that bar in Bangkok! If Ted had succeeded in all of this, then the promise of BNF of 40 years could now be realised but not by root nodules and Rhizobia, or through molecular manipulation but by a different paradigm altogether!

2

AN EXTRAORDINARY SCIENTIST AND SCIENTIFIC PHENOMENON

Nullius in verba

I have personally always been a little bit wary of being a member of any organisation since membership always seems to me to imply patronage, subservience or belief in a doctrine to which I may not at all times subscribe. I would never want to be hypocritical or placed in a position where my integrity is called into question, so I tend to keep memberships to a minimum. I have however, always had the greatest respect for those institutions whose core values have been demonstrated to hold true over historical time. In the UK there are of course a number of long established institutions that may rightfully make that claim, but as a scientist, the Royal Society stands pre-eminent in this regard.

The Royal Society was established in 1663 as the *Royal Society of London for Improving Natural Knowledge*. Today The Royal Society is the UK's national science academy with a Fellowship of around 1,600 of the world's most eminent scientists. The Royal Society's motto *'Nullius in verba'* is taken to mean 'take nobody's word for it'. It is an expression that is intended to reflect the determination of The Society's Fellows to withstand the domination of authority and to verify all statements by an appeal to facts determined by experiment.

I have always been intrigued by that part of the motto "...to withstand the domination of authority..." because to me this includes the courage and ability to stand against the prevailing orthodoxy; the ability to question and drive for paradigm change. And in fact this is exactly what so many of the Fellows of the Royal Society have achieved over the institution's illustrious history; Fellows have challenged the existing orthodoxy or paradigm and supplanted it with another that better relates to evidenced-based knowledge.

Professor Edward Charles Daniel Cocking was elected as a Fellow of the Royal Society in 1983; a year of Royal Society elections that was quite notable because it honoured the UK natural historian Sir David Attenborough and by Statute 12 also the Right Honourable Margaret Thatcher MP the then, rather controversial Prime Minister of the UK. Ted's appointment like the former was highly deserved and certainly unlike the latter caused no controversy because he had made, even by then, an outstanding contribution to plant science.

If one lists the qualifications of an individual, then these can look impressive, and this is certainly true of Ted Cocking. In such a formal CV style, Ted received a BSc in biological chemistry from the University of Bristol in 1953, followed by a PhD and a DSc in plant cell biology and nitrogen metabolism. After receiving a three-year Bacterial Chemistry Research Fellowship, he was appointed Lecturer in plant physiology at the University of Nottingham in 1959. He subsequently became Reader then Head of Department and Professor of Botany at Nottingham. These are all significant achievements but they do not really begin to help one understand or to explain the man, or the research he was undertaking and how it set the scene for later achievements; achievements that form the core of this narrative.

Perhaps the early post-WWII start of his scientific career working on bubonic plague vaccines at Porton Down, the UK Microbiological Research Establishment, which were used to vaccinate US soldiers in the Vietnam War, gives an indication of his potential. At that stage there was no indication that he would go on to define techniques for isolating plant cells (called protoplasts) enabling both inter-plant and plant and animal cell fusion, that made possible subsequent developments in biotechnology and genetic manipulation. However, it was not many years before his potential future impact was noted. As early as 1971 in an article entitled "The New Botanists" published by Graham Chedd in the *New Scientist* 29th April 1971, Ted Cocking was already creating a stir as a young Professor. Chedd writing about Ted Cocking's research group stated "... the Nottingham work pivots on single, naked plant cells, which the group is learning to manipulate in the manner that animal cell biologists have been doing with their working material for some years. The difference is that from the plant cell work will probably emerge an agricultural revolution". Would such a prophetic statement turn out to be justified?

A Naked Cell

Animal and plant cells are similar in that they each have a cell membrane, filled with a gel-like substance called cytoplasm and a single nucleus and organelles, of which one of the most important are the mitochondria that allow cells to respire. However, plant cells differ in that they also have a rigid cell wall (enclosing the cell membrane), a vacuole inside the cytoplasm (a space that fills with sap to keep the cell turgid) and one or a number of chloroplasts (a structure containing green chlorophyll to absorb light energy to enable plants to photosynthesise). The cell membrane maintains the integrity of the cell but even so it is porous and is responsible for regulating what moves in and out of the cell. The membrane is essentially a double layer of lipids (fat is an example of a lipid) embedded with proteins and carbohydrates that create channels through the membrane and enable cells to communicate with each other.

The additional rigid cell wall in plants is essential to provide mechanical strength and support and to withstand the turgor pressure exerted by cell contents that keep the plant rigid and erect. The cell wall helps retain water and we have all witnessed how plants wilt in the absence of water due to this lack of cell turgor. The cell wall gains its strength from a substance called cellulose, which can retain flexibility in young cells that allows growth, and yet, becomes more rigid as it fills with lignin in older cells. Although the cell wall is porous to some substances it also acts as a barrier to protect cells against attack from plant diseases caused by bacteria, fungi and viruses. In addition, it also acts as an effective barrier against scientists wishing to get to grips with a deeper understanding of internal plant cell function.

Ted was trained as a plant biochemist and was one of many scientists frustrated by the difficulty of having to work with whole plants simply because it was not really possible at the time to work with individual or groups of plant cells. The only method available was to physically remove the cell by mechanical methods that cause some cell damage and therefore limits what can be done. Mechanically extracted plant cells certainly won't undergo further cell growth or development. Ted was frustrated by this inability to readily produce viable individual plant cells, a frustration piqued by a research project in which he had already published two significant papers. In this project he had been working with bacteria. One of the great things about working with bacteria is that they can be produced in simple culture in a matter of days so that single or colonies of cells are available and amenable to biochemical study. Ted therefore started to look for ways of isolating individual plant cells so he could work with them in a similar way to bacteria. His first target was the rapidly growing region of the roots of tomato seedlings and the outer layer of the plant cell wall that he tried to disrupt in order to separate and isolate individual cells. In his first attempts Ted used substances called chelating agents for this purpose, and although these provided him with the separation of individual cells that he sought, he could not then achieve any division of these cells to produce a useful colony. Unfortunately it was evident that the chelating agents he was using were damaging the cell physiology and biochemistry of the cells. A different approach was needed.

Progress in science is often limited by the constraints of a particular methodology, and a research discipline at any one time may even be defined by the limitations of the methodologies at its disposal. A new methodology or experimental technique can cause a vertical leap in progress - opening up whole new opportunities and areas of study. New methodologies can literally be completely novel approaches or involve the transfer of tried and tested techniques from one discipline to another, but in either case they can transform what's possible within a discipline. And so it was for Ted and his failure with the chelating agents. Ted looked elsewhere for a solution, and on this occasion not just a technique that would separate cells while keeping the plant cell wall intact, but rather a technique that could remove the cell wall altogether - techniques that would produce naked plant cells. The technical name for such naked plant cells, are 'protoplasts'.

Ted had read about the isolation of individual bacterial and fungal cells (which like plants have outer cell walls) through the use of enzymes called lysozymes that degrade specific molecules in the cell wall of the bacteria. Ted's training as a biochemist came into play as he realised that lysozymes would not be appropriate for plant cells, but enzymes called cellulases that degrade cellulose might work. Taking the principle of use of enzymes from microbiology, Ted applied it to a similar problem in plant science - and he did so with great success - but not immediately.

First Ted decided to survey a wide range of commercially available cellulase preparations to see if any would break down the plant cell wall and release a naked plant cell. Using seedling roots which are actively growing and hence, at any one time have cells at many stages in their development, the different enzymes were applied and tested - but all without success! Ted was surprised by this and not a little perplexed but with his usual tenacity, he reviewed what he knew. It was then he remembered a piece of research by D. R. Whitaker at the National Research Laboratories, Ottawa, Canada on the purification of cellulase from *Myrothecium verrucaris*. This is a species of fungi that causes diseases in plants and is often found on materials such as paper, textiles, canvas and cotton. *M. verrucaris* is known as a highly potent cellulose decomposer. In November of 1959, Dr Whitaker had generously provided Ted with a small sample of his enzyme preparation that Ted had safely filed away in the bottom of a freezer in his laboratory. Only when all else had failed and Ted was searching for anything that would meet his needs did he suddenly remember the sample from Dr Whitaker. Ted rummaged in his lab freezer and used some of the sample as an application to a root preparation; and to his relief and great pleasure, 'Eureka' the enzyme released the plant cell protoplasts from their cell walls. Ted had discovered a new technique for generating naked plant cells! It was one of those rare but crucial moments in science that subsequently ushered in a new era of plant cell research, but of course at the time no one realised just how significant this would be!

The Benefits of Naked Plant Cells

With the achievement of producing plant protoplasts came the realisation that his laboratory had no means of scaling up their capability. Insufficient enzyme was available for anything other than attempts to isolate protoplasts from roots which held-up extensive large-scale experiments with leaf protoplasts. In another case of serendipity Ted's team had by accident discovered in 1963 that by using a commercially available enzyme, pectinase, it was possible to isolate protoplasts from tomato fruit because the particular site they targeted had cell walls made of pectin. They then developed the ability to produce large quantities of these isolated protoplasts throughout the 1960s. This enabled Ted and his team to undertake experiments exploring the properties of, and the ways to, manipulate the plasma membrane in the absence of a plant cell wall. Ted was very interested in how the cell wall could regenerate but also, and important for his later breakthroughs, he studied the process by which the naked cell can interact with, and actually take-up, different types of particle. This is a process known as endocytosis, which involves the plasma membrane engulfing the particle and enclosing it within the cavity this creates. This process of engulfing or invagination to create a cavity (known as a vacuole) provides a means for materials to pass into the cell. In this context, Ted and his team started to look at how viruses that attack plant cells actually infect and get inside the plant cell. They worked with the tomato mosaic virus (TMV) and using electron microscopy demonstrated that endocytosis is the means by which the virus enters the plant cell. The virus is enveloped by the plasma membrane so that it is completely surrounded which then turns in on itself to form a pouch containing the virus within a small fluid filled vesicle that is inside the cell.

This demonstration was important because it essentially established for plant virologists a model system whereby they could simultaneously infect a number of cells with a virus, and because they had access to a high proportion of infected cells, they could carry out meaningful biochemical and molecular investigations. For Ted it opened up a means by which a whole range of different materials could be inserted into plants cells and studied, whether that was DNA, micro-organisms or organelles.

In 1968, the cellulase from the fungus *Trichoderma viride* was being produced in Japan for baby food and biscuit manufacturing. The enzyme was then available commercially for Ted to buy thereby making possible the more widespread study of protoplasts and more scientists exploring what could be achieved with basic plant cells. With animal cells it had been possible to multiply cells in tissue culture and the goal, having separated individual plant cells, was now to find a way to regenerate whole plants from such cells. If this were achieved it would generate new opportunities of (i) taking cells from different plants, fusing them and growing a novel combined (hybridised) plant and (ii) inserting DNA or micro-organisms into plant cells to produce transformed plants having novel traits and attributes.

The first step towards regenerating whole plants from individual cells was achieved in 1969 by Ted and his team, involving the fusion of individual wheat and barley protoplasts. However, while the cereal leaf protoplasts would fuse and readily re-synthesise cell walls they would not undergo sustained cell division and grow into whole plants. This was first achieved using tobacco leaf protoplasts by the pioneering work of George Melchers, George Labib and Itaru Takebe that was published in 1971.

There are a number of other key developments that occurred in the early 1970's that are important to our story. The first of these is that Ted was working with a number of enzymes, one of which was Beta-1,3 glucanase that was being used in a mixture to improve the isolation of the protoplasts. This enzyme is worth a quick mention because it has an important role later on. The second development was less esoteric and that was the offer of funding by Arnold Spicer, the then research director of the Lord Rank Research Centre. Ted was offered funding to investigate, with a great deal of autonomy, any novel aspect of protoplast biology that he wished! This is not the sort of open offer that occurs in the present day, but those were different times and it was an opportunity not to be ignored. The subject Ted chose to study was the uptake of bacteria by plant protoplasts. The question Ted wanted to answer was whether or not it was possible to introduce the nitrogen-fixing legume bacteria 'Rhizobia' without the need for root nodules, into non-legume plant cells.

Interestingly although Ted's initial attempts at introducing Rhizobia into non-legume plant leaf protoplasts demonstrated that it was possible in principle for the bacteria to be engulfed by a protoplast, he was unable then to achieve regeneration of the whole plant. Mike Davey in Ted's group decided to look at the isolation of nodule protoplasts containing Rhizobial bacteroids and whether their fusion with non-legume leaf protoplasts could be achieved, but again they could not obtain any regeneration of the non-legume plants containing intracellular Rhizobia.

Although in scientific terms these experiments did not achieve the desired result of nitrogen-fixing Rhizobia distributed throughout a non-legume plant Ted was able over a number of years to learn from these failures. Such knowledge formed the basis on which he was to define the attributes required for bacteria to colonise plant roots - one of which was the presence of cellulase enzymes such as Beta 1,3-glucanase, mentioned earlier.

Taking a different tack in the 1980s Ted decided to investigate the possibility of the direct interaction of foreign DNA with plants, with the objective of eliminating the need to isolate protoplasts. He wanted to subject the protoplasts to the various manipulations required for the introduction of DNA from non-plant sources and then to spend time regenerating whole plants from the modified protoplasts. However, what appeared simple in concept was not so easy in practice.

After several years of effort, Ted was able to describe an enzymatic procedure for the degradation, within a few minutes, of the cell wall at the tips of root hairs from a wide range of crop species that exposed the plasma membrane with partial protoplast release, whilst maintaining the functional integrity of the plant. Such exposure of the plasma membrane provided a point of entry for Rhizobia. Sadly however, although such enzymatic treatment of legume root hairs was shown (i) to remove a barrier to Rhizobial-host specificity in non-legumes, and (ii) to enable the formation of nodule-like structures on rice and other non-legume crops, the ultimate goal of nitrogen-fixation within these nodules on non-legume crops, did not result.

This progress, although not wholly successful had along with the research from other groups, raised awareness of the potential of nitrogen fixation through the use of bacteria in crops other than peas and beans. The Rockefeller Foundation then invited Ted to organise a discussion meeting with Ivan Kennedy from the University of Sydney, on the subject of *Biological Nitrogen Fixation: The Global Challenge and Future Needs* (Cocking, 2017). From wide-ranging discussions at the meeting, the importance of trying to establish diazotrophs, such as Rhizobia, within the root systems of the world's major cereals was highlighted. It was now 1997 and the prospect of utilising genetic manipulation techniques to generate root nodules producing capability in non-legumes was being pursued with vigour by a number of research groups. The manipulation of protoplasts as a route to production of Rhizobia within non-legumes by Ted, his team and like-minded individuals, was beginning to run out of steam. The future at that time was very much seen to be reliant on the need for genetic manipulation of the plant to produce root nodules that could then be colonised by Rhizobia. Ted's future direction of travel was about to change again and it did not involve the dream of transgenic nitrogen-fixing plants!

Gluconacetobacter diazotrophicus

Bacterial species names are often difficult to pronounce. *Gluconacetobacter diazotrophicus* is no different but the easiest way to think about it is phonetically - thus: *Glucon - aceto - bacter diazo - trophicus*. However, just to confuse things a little further when the bacteria was first identified it was given a slightly different but simpler name to pronounce - *Acetobacter diazotrophicus*. Such is the world of science and taxonomy that although one might become attached to particular species' names, they can change on the basis of revision of the phyla by taxonomists. However, whichever of the two names you come across, they are synonymous and we are talking about the same bacteria.

Rather than require me to write out the full name or for you to read it, you will I hope be satisfied with the use of the following abbreviation 'Gd'. Not long after it was first discovered, Gd was described as an "extraordinary endophyte" (endophyte meaning within a plant). What has become clear over the last 30 years is just how extraordinary this endophyte really is!

Gd was discovered in 1988 by Joanna Döbereiner and Vladimir Cavalcante (Cavalcante and Döbereiner, 1988) in sugarcane in Brazil. The bacterium was discovered and identified during the 1980's simply because the Brazilians were looking to increase production of ethanol from sugarcane to use as a substitute for petrol. While investigating the means of improving sugarcane production it became apparent that some fields of sugarcane had remained unfertilised for decades, but they did not appear to have suffered any yield loss. This was extraordinary and suggested that biological nitrogen fixation might be occurring. Joanna and Vladimir started to sample the sugarcane juice in search of suitable candidate bacteria. The bacterium they isolated from the sugarcane juice was *Gluconacetobacter diazotrophicus* and they were able to demonstrate that indeed Gd was able to fix nitrogen. The isolate from Brazil was described in detail by the means available at the time and it became the international reference strain for the species, against which all others have subsequently been compared. It was originally labelled as PA15, although intriguingly over time this has mutated into a general use of the term PAL5, and this is the name now regularly bandied about by normally very precise and pedantic scientists when referring to the reference strain.

Döbereiner and Cavalcante's discovery of the bacteria from sugarcane set other scientists onto the trail of Gd. One scientist in particular deserves specific mention in this regard - the late Dr Jesus Caballero-Mellado, an academic at a university, UAP in

Mexico - now the National Automonous University of Mexico (UNAM). Caballero-Mellado isolated a number of strains of Gd from sugarcane from around Mexico, and was quite happy in the spirit of international scientific collaboration to send upon request isolates to other scientists around the world. One such request came from a Professor Ted Cocking at Nottingham University and a package of isolates for academic use was duly dispatched to the UK. Ted Cocking received these in 1998. That package contained a strain of Gd referred to as UAP5541.

At the time Ted was screening bacteria that would naturally express enzymes, in particular Beta -1,3 glucanase or similar, in order to break down plant cell walls - to produce naked cells whereby intracellular colonisation of the bacteria might occur. So when the bacterial strains arrived from Professor Caballero-Mellado Ted set about screening for these attributes. In UAP5541, Ted was delighted to have hit the jackpot!

Ted found that if he provided the Gd strain with a source of energy in the form of sucrose and inoculated only a few bacteria into an agar medium around the roots of growing seedlings, he was able to achieve intracellular colonisation of Gd of the roots. Ted tested this with a range of plant species, with the model plant *Arabidopsis thaliana* and the crop plants maize (*Zea mays*), rice (*Oryza sativa*), wheat (*Triticum aestivum*), oilseed rape (*Brassica napus*), tomato (*Lycopersicon esculentum*), and white clover (*Trifolium repens*) and achieved the same result, intracellular colonisation of Gd in the root cells. Ted was able to use a well-established technique to produce what is known as Beta-glucuronidase (GUS)-labelled and with NifH promoter–GUS-labelled *Gd*; a rather complicated name for a coloured tag that means they show up blue under a light microscope. In his studies of the root tips of the inoculated plants Ted was able to show the presence of the blue-stained *Gd* within the cytoplasm of root cells, indicating that intracellular conditions were suitable for nitrogenase gene expression. Electron microscopy confirmed that these blue-stained intracellular bacteria were within membrane-bounded vesicles.

Such membrane-bounded vesicles have a specific name, they are referred to as a 'symbiosome', and their presence is significant because they are generally regarded as a necessary structure to truly enable symbiotic intracellular nitrogen fixation. Ted's colleague produced a schematic of the process (Figure 2.1). Ted hypothesised that the release of enzymes allowed the bacteria to punch tiny holes in the outer skin of the root and enter the root. Once the Gd had entered the root tissues the bacteria is able to move through the plant by a number of different mechanisms, simply as a physical process moving with the flow of nutrients and water in the xylem vessels or more actively through reproduction in rapidly dividing plant cells so that as the plant grew, so the cells developed with Gd inside them. The Gd could enter individual plant cells through further release of enzymes that break down the cell wall and through a process of endocytosis, which Ted had first observed with tomato mosaic virus, leading to the encapsulation of Gd within symbiosomes within the plant cell (Figure 2.1).

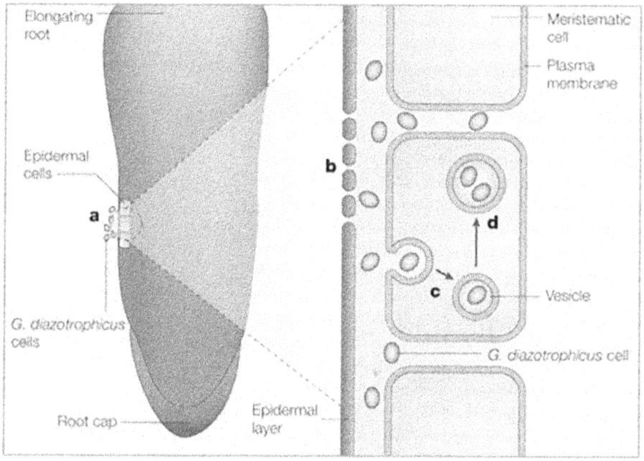

Figure 2.1. Schematic diagram of the proposed mechanism of colonisation (a) by Gd on the root exterior releasing enzymes to punch holes through the root epidermis (b) and the process of endocytosis (c) that facilitates entry into the cell and formation of a vesicle like symbiosome (d) in which the bacteria can multiply. (Reproduced by kind permission *ACSESS 2006* ©).

Using the blue tagging for the bacteria Ted was able to demonstrate the Gd (black dots in Figure 2.2) moved through maize plants to the leaves, and could be found next to the chloroplasts (circular-disc like structures) that are responsible for fixing carbon dioxide from the air to produce sugar for the plant (Figure 2.2). Gd needs sucrose as a source of energy, so it is perhaps not surprising that the bacteria accumulate around the chloroplasts where the highest concentrations of sucrose may be found.

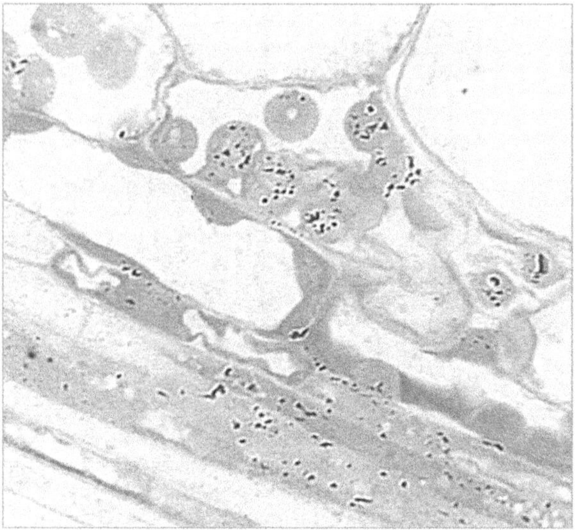

Figure 2.2. Image of Gd (black dots) inside the leaf structures including the leaf chloroplasts (circular disc like structures). (Reproduced by kind permission *ACSESS 2006* ©).

Ted's research with Phil Stone went on to demonstrate, using a technique called Acetylene Reduction Assay (ARA), (a well established and commonly used but not wholly reliable method) that Gd while colonising intracellularly was also fixing nitrogen!

Realising the commercial potential of such a finding Ted and the University of Nottingham filed a patent for the conditions under which plant intracellular colonisation of Gd could be achieved in a range of crop plants and deliver nitrogen fixation. Subsequently Ted, Philip and a colleague Mike Davey published a scientific paper on this development (Cocking *et al.*, 2006).

3

FROM A MEETING OF MINDS AND ... MEN?

Returning from Thailand

Having returned from the Thailand Workshop I had immediately contacted Alan to follow-up on our Bangkok bar conversation. Not long after on the 21st March Alan sent me a University of Nottingham confidentiality agreement (a CDA), sometimes called a non-disclosure agreement (an NDA), whereby I would commit to not disclose any information I was to learn to another party unless authorised to do so by the University. This agreement allowed me to talk freely with Ted to gain a better understanding of the detail of his scientific research and to have access to information that was not publicly available.

Since my return from Thailand I had also been in contact with Dr John Beddington, the Chief Scientific Adviser to HM Government, to update him on the progress of the Thailand workshop since the idea for this had arisen out of a visit John had made to the country the previous year. At the time I also mentioned Ted's work on nitrogen fixation and since John was visiting Nottingham, I suggested that he enquire about this research during his visit - which he duly did.

A Noisy Café in Nottingham

Four very different men met at the Lakeside Café on the University of Nottingham Park Campus on 13th April 2011. Dr Alan Burbidge whom you already know but deserves a more detailed mention, Dr Phil Stone, a key character in the next few years of this story, Prof Ted Cocking of course and myself.

Alan is one of those too rare individuals who is self-effacing, modest (both attributes of which tend to be in short supply among scientists), intelligent (there are many such), not self-serving and commercially astute (not so many at all) but most of all, a man with integrity! I leave it to your own experience of life to judge the value and scarcity of the latter attribute. But in fairness, I did not know him well enough at that the time of this meeting in April; it is something I came to understand and appreciate over time. What I did know at the time was that as Licensing Executive for University of Nottingham he was enthusiastic and committed to commercialising inventions from the University of Nottingham; Alan is a 'fixer' - he made things happen.

Alan had a background as a scientist in plant molecular genetics and before moving in 2003 to the role of Business Development Executive he had been a Postdoctoral Research Fellow at the University. As the relevance of innovation became increasingly important a new role of a Licensing Executive for Life Sciences technologies in the Department of Business Engagement and Innovation Services was created. Alan's job was to assist academics such as Ted Cocking, to commercialise their inventions through promoting collaboration, suitable partnerships and agreements. For Alan this meeting between Ted, Phil and I was a first step in a sequence that hopefully would help him find a suitable commercial partner to exploit the patents the University had filed on Ted's work.

The initial introductions in a public place for any group of people, who do not know each other well, are often awkward - and this was no different. Ted's scientific reputation and my own hopes and expectations of the meeting certainly made me feel a bit on edge, and I wanted to make a good impression. The bustling activity and chatter of the Lakeside Café was not going to make this easy for any of us. Phil and Ted came in together and they shuffled between the tables and chairs, over to where Alan and I sat, camped around the end of a long table.

Anyone who has met Ted knows that he is a large silver haired man with a certain physical 'presence' that he retains still from his youth, now though with a slight stoop indicative of his advancing years. Ted was dressed as was I, in traditional business attire - he was self-confident but somehow unassuming with a quiet almost whimsical and rather disarming demeanour. This highly distinguished academic contrasted dramatically with Phil Stone his postdoctoral scientist, who seemed jittery and ill at ease. A smaller man with long dark hair that hung limply to his shoulders that combined with a rather casual, dishevelled appearance that initially did not inspire me with a great deal of confidence; a view I am pleased to say that was mostly dispelled in the next hour of conversation and then in subsequent months as I got to know him well.

As Alan introduced us, I was initially thrown off guard as I noticed what everyone notices in their first introduction to Ted, that he has an eye condition known as exotropia. This can be disconcerting as one attempts to determine which eye to follow when meeting face-to-face. It was certainly disconcerting for me, as I struggled too hard not to fixate, to relax and to shrug off the informality of the circumstances. None of this was how I had imagined it would all go - this was not what I was used to - professionals in professional attire, in a professional meeting room discussing things professionally! And it proceeded to get worse.

I explained my background, expertise and experience and how I thought I may be able to play a part in commercialising Ted's work, as the Lakeside café filled with mums and toddlers and other visitors, and the noise level continued to rise. As I started to listen to Ted's and Phil's explanation of their research it was a bit like the broken conversation on a mobile phone struggling to maintain a signal, I could not hear every word spoken and it was increasingly frustrating. What I did hear was just sufficient however, to gather the gist of each topic covered by Ted and to enable me to ask vaguely relevant questions in turn. We struggled on bravely in this vain for nearly an hour until we all realised that enough was enough. I had heard sufficient to know that the prospect of working with the University and Ted to identify opportunities for commercial exploitation of their technology excited me greatly, and it seemed that they had learned sufficient about me to have confidence I could deliver something of relevance for them. We left the confines of the café to the relative quiet of the surrounding park and said our farewells. Although the circumstances of our meeting had not been ideal we all had agreed the next very positive steps.

The first of these was an action to be taken by Alan who formally commissioned my company Dent Associates Ltd to undertake a consultancy on behalf of the University of Nottingham to investigate the commercial opportunity on offer and to identify potential licensees for their nitrogen-fixing technology. Alan had been working hard to keep the technology under development and active even in the absence of a licensee, through successful applications for internal University funding, but now appropriate commercial partners were needed and he hoped Dent Associates could identify these. Alan had previously thought he had found suitable partners. He and Ted had engaged with a specialist consortium and subsequently with an international corporate but both had failed for different reasons. In one case this was because the licensee had not listened to technical advice from Ted, because they thought they knew better rather than learning from the years of experience of working with Gd that Ted had accumulated. A licensee was required that could build upon existing knowledge and work with Ted Cocking utilising from the outset well-honed methodologies and protocols.

At that time, the University had a keen interest to look at commercial opportunities in Brazil, not least because there was a track record of research and commercialisation associated with Gd and other nitrogen-fixing bacteria. Brazil then became an obvious starting point and a focus for the research we at Dent Associates were to undertake. I undertook the project with Rogerio Ghesti, a Brazilian national and one of my consultants at the time, between April and May 2011. Thankfully during that time I was able to meet with Ted and Phil Stone on a number of other less fraught occasions to ensure I fully understood their technology, the limits of their current research and their expectations for its exploitation. In carrying out the analysis my company was able to identify routes to market including potential partner companies with an interest in licensing the technology but I also got to grips with some of the barriers to commercialisation and what regulatory matters might arise. Most importantly of all however, we began to understand the true scale of the opportunity and market need for a seed treatment that could genuinely provide an alternative to conventional nitrogen fertiliser use. This market was immense and if a seed treatment could be even moderately effective the impact for agriculture could be enormous. Dent Associates formally completed its assessment and submitted its report in June 2011.

It was now I started to think that I needed to find a way to assist further in the commercialisation of the Ted's technology. Here was something that could potentially revolutionise agriculture if it could be delivered and distributed globally. I realised then, that I wanted to commit to playing a key part in making that happen.

From Concept to Company

I had been running Dent Associates as a successful consultancy business with a small number of part-time consultants for 6 years. I was doing well, I worked from home, I chose my clients and projects, and travelled internationally occasionally to countries of my choosing. I really had no need to change my lifestyle or to take any risks; life really was just fine. But, the truth of the matter for me was that it was too easy and I needed another challenge. Life had not been dull by any means. During the time I had built-up a client base with Dent Associates, I had also invented and patented the technique for rounding-up transactions via PIN terminals for charitable causes and founded the Pennies Foundation, I had developed my proprietary analytical technique for identifying market-gaps for innovation as well as having published a report with Professor Mike Theodorou on 'Market-led Innovation Centres; a new model for UK innovation'. After meeting and corresponding with Vince Cable MP, this report fed into the Coalition Government's Research and Innovation Strategy. So it wasn't as if I hadn't achieved anything. However, all this paled into insignificance compared to the potential value to society if it was possible to translate Ted Cocking's laboratory proof of concept into a working commercial product!

One of the things I had learned while setting up the PIN terminal rounding-up "Pennies for Change" scheme is that it is easier to be seen as credible and to leverage influence by engaging with the right partners. The range of skills and expertise required to make a success of an enterprise does not all normally reside in a single person. I have always felt I have a good sense of my own strengths and weaknesses and the sheer effort that would be required to make a success of such a venture. This nitrogen-fixing project was going to be bigger than I could manage alone. I needed partners! Of course it is always true to say the initial choice of partners may not always best suit the opportunity further down the line, but no one has a crystal ball and so to some extent it is always about making pragmatic decisions at the time they need to be made. In this regard, a pivotal step in this story was an email that I sent on 23rd August 2011 to Peter Blezard, the former CEO of Plant Impact plc, who having left the company was looking for a new project.

I had known Peter from my time as Managing Director of CABI Bioscience, when he as CEO of Plant Impact plc contracted my organisation to carry out international field trials on some of the company's products. On leaving CABI, I had been employed as a consultant to Plant Impact to chair the company's Scientific Advisory Board, and had subsequently met Mike Panteli (Chief Finance Officer) and Allen Sheena (Marketing Consultant who ran his own consultancy business). Peter and Mike had finished with Plant Impact and Peter was at a definite loose end. He had been in contact with me on a number of occasions looking for, and enquiring after, any new project opportunities of which I might be aware relevant to his talents and skills. My email on the 23rd August to Peter described Ted's intracellular colonisation and nitrogen fixation invention and it's potential. I described it thus, "If you want a new technology to work with in agriculture - this is potentially revolutionary ..."

A follow-up phone call on the Friday of that week started the process rolling for what would eventually become a new company and the development of a new nitrogen-fixing technology. I contacted Alan Burbidge and enquired whether the University would be amenable to licensing their patent to a new company that would secure investment, develop and commercialise the technology. Having got to know me well, and trusting my judgement Alan was prepared to have a conversation to discuss what we had in mind.

In November 2011 both Peter and I presented a proposal to Alan Burbidge for a NewCo. Hemsley (based on the name of the meeting room at the University in which Peter, Alan and I were to meet), a company to license and commercially develop the technology based on the patent of Ted's invention. The meeting was very positive and over a number of meetings with Susan Huxtable Director of the University's Technology Transfer Office and the University lawyers, NewCo. Hemsley morphed into Azotic Technologies and was formerly registered as a limited company in January 2012. Following a number of months of negotiation the Company secured an exclusive global license from the University of Nottingham for the commercial exploitation of Ted's nitrogen-fixing technology.

And why 'Azotic'?

Naming a new company can be a contentious issue with every one of a team having their own favourites. With Azotic Technologies, I provided a basic rationale for a set of potential names associated with Greek and Latin words meaning 'nitrogen'. When the French chemist Lavoisier discovered nitrogen he called it 'Azote' meaning 'without life', because of its inert nature. Hence for me, with nitrogen-fixing bacteria known as diazotrophs, derived from the word 'Azote', it was a simple leap of imagination to the name Azotic and hence, the company name - Azotic Technologies Ltd. We threw around a number of other ideas but could not come up with a better rationale or name and so it stuck - we were Azotic Technologies Ltd or if the telephone line was poor 'Exotic Technologies', which was a source of amusement on a number of occasions! For the most part Azotic Technologies was a name to which most in the industry could relate and respect and throughout this narrative I abbreviate the full company name to just 'Azotic'.

Tentative First Steps

Coming up with a name is the easy part of starting-up any company. Those first few months of organising the fun stuff; names, logos, titles and business cards, website and new email accounts - everything and anything in the world seems possible! However, running a business is a serious and heady responsibility. If the Company is to gain any traction and grow then, there are many more down to earth, operational issues and processes to set in train and to make work.

The standard advice is always to prepare a business plan, and while this can be rationalised as a meaningful and useful exercise, its real value is in forcing all those involved into organising and preparing the ground for all the hard work that is going to be required in order to make a company a success. It is actually quite difficult to have the necessary foresight, knowledge and experience to ensure everything you need to address is written into a business plan. This is simply because as things develop, all of the assumptions that have been made at the start will be challenged and a good number of them will prove to be irrelevant or even wrong. There will be times when even those who genuinely wish to help, and certainly all those who want and even enjoy doing so, will put a spanner in the works for you! You have to work out the difference between what is good advice, what you currently know and believe and those who criticise and seek to undermine you with information counter to your plans. Not everyone will want you to succeed, especially those that are potential competitors in technology development or in the market place, and you will also inevitably receive negative feedback from potential investors simply because they want to decrease your share value prior to proposing an investment. A balance has to be achieved between stubborn commitment to your expertise and beliefs or modification to meet a genuine need for a fledgling company.

Tenacity is an absolute requirement for individuals in any start-up. Even so it is important to understand that the messaging, product ideas and markets will change - or rather, they will evolve as you explore with potential investors and partners the true nature and scale of what you have to offer. This is a process of development that must be embraced even though it will test everyone within the team whether from a technology, regulatory or a commercial perspective. Without a doubt the initial simplicity of a business concept will be challenged from every conceivable (and some inconceivable) angle, as different, widely varying levels and types of expertise are brought to bear in an attempt to expose weaknesses in your plans. There is usually nothing malicious in this but rather the need to determine the competence of the team and their ability to cope with criticism and challenge, but also to assess the thoroughness or otherwise, of the technology and the Company's business plans.

Initially, this process of challenge will expose areas and issues you may not have considered, have inadequate answers for or just cannot know at such an early stage. However, if you thoroughly investigate and address the issues so that your knowledge and understanding grows in line with your ability to communicate these, then it becomes easier to address and resolve the misgivings of potential investors and other critics. The ultimate winner in this process is your business, but the process in itself can be quite demanding. I always likened an investor interview as the equivalent of a PhD viva, where you are questioned and in some cases interrogated for a couple of hours, during which time, everything you know is challenged, your work, its value and even your own motives are questioned. It can be exhausting - and for a Chief Technology Officer (CTO) with a potentially disruptive technology, one is up front and centre stage - time and time again. As the expression goes, you will need to "Gird your loins" for this process - it can be unrelentingly tough!

During the process of engaging with potential research and commercial partners, key influencers, investors, regulatory bodies and government then every aspect of your technology, business and markets will have been scrutinised and evaluated from a myriad of different perspectives. Surviving and emerging triumphant from such a process is more involved and detailed than any other process of 'peer' review of which I am aware. Certainly from a scientific and technical point of view the idea often espoused that peer-reviewed scrutiny of research for publication in a scientific journal is somehow more rigorous is, in my opinion, simple nonsense. Any technology that makes it through to the market place has been thoroughly assessed and appraised almost to destruction, and finally will face the ultimate challenge of scrutiny by customers - which is a continuous and rigorous process in its own right.

This scrutiny of the Company has additional benefits since it will highlight the adequacy and skills of the management team, identifying where resources may need to be dedicated to address gaps in capability. Failure to learn in this regard leads only to problems that will accumulate as the company grows. It is my belief that as an individual you need to grow with your company, or leave it to others to do so. This can be particularly difficult for 'inventors' but it is why investors will sometimes suggest building a team around the originators of a technology or company.

In our case, having spent a year building the case for Azotic and Ted Cocking's technology, Peter Blezard secured investment from an Angel Investor in North America that would give us sufficient funds for a single year; one year in which as CTO, I needed to translate a laboratory based concept into a technological product that could be evaluated in the field and generate results that would be sufficient to justify further investment in the Company. To say that this was a tall order completely misses the point; yes there is risk and anyone who has any entrepreneurial spirit is used to dealing with the pressures of risk. But the other side of the coin is 'responsibility'; for me it was about having been granted responsibility for a technology that had so much potential, a disruptive technology, developed by an eminent scientist, whom I admired and respected and who in turn trusted me to deliver for him. In my mind it would have been irresponsible to forget that it was Ted who was placing into our hands an opportunity to deliver on his personal vision and hopes for a revolution in sustainable agriculture. And to this day I have never forgotten nor reneged on that responsibility.

The Green and Greener Revolution

There have been many notable scientific advances over the last 100 years and some of the men behind them (and for right or wrong they have been mostly men) have received well-deserved accolades for their achievements. In agriculture, one man stands head and shoulders above the rest for an achievement with global implications - that man was the late Norman Borlaug who received the Nobel Peace Prize for driving the Green Revolution. Dr Borlaug is credited with saving more lives than any other person who ever lived. He did this through developing a high-yielding short-strawed, disease resistant wheat which he then led into extensive production, bulking up and releasing to farmers globally. The availability of these revolutionary varieties to farmers ensured that those most in need particularly in Asia were able to produce high yielding crops, thereby averting widespread famine. In his own words "a temporary success in man's war against hunger and deprivation". Vast areas of the new wheat were sown worldwide with the major beneficial impacts on yield creating what became known as 'the Green Revolution'.

I met Norman only on one occasion. Having been a Member of Council of the Parliamentary and Scientific Committee since 2002, and during one of my seven years of service as a Vice President of the Committee, the then President Lord Soulsby of Swaffham Prior and the Chair Dr Doug Naysmith proposed inviting Norman to address the Committee, which he did on October 26th 2005. I was lucky enough to be one of the few dinner guests after the event and to have the chance to speak to Norman. He was a truly remarkable man; quietly spoken unassuming and a very polite gentleman who sat in conversation with me as a living embodiment of all it should be to be a man who has achieved truly great things.

Whereas I had the privilege of meeting Norman on only one memorable occasion, Ted on the other hand knew Norman well. Norman had often lamented on the need to replace nitrogen fertilisers with a sustainable alternative and had actively encouraged Ted in his endeavours to produce a nitrogen-fixing technology for cereal crops. Norman had fostered Ted's vision of nitrogen fixation being ubiquitous among all crops as a means of addressing the world's food security and climate smart agricultural needs - a 'greener revolution'!

I think it is always important to ask yourself that if by chance, you had an opportunity to make a positive difference in the world, even a small one, but in doing so it would involve going out on a limb, taking a bit of a risk - would you take it? Would you take the risk, would you bear the cost, the potential ridicule and questioning of your peers, and the prospect of failure! As I embarked on this journey, I only had to look for the answer to this question in the lifelong commitment and drive of one man, of Ted Cocking. Here was a man with a long held vision; a vision consistent with that of Norman Borlaug, with the needs for keeping pace with feeding a growing population, a more sustainable approach, that also would start to address climate change issues in the field of agriculture, my field of expertise. I recognise to this day, that there are not many of these types of opportunity that ever come along. And I had embarked on one of them. I also knew that if Ted's vision could be realised, - if it could be done, then I was convinced it had to be done, and I would do my utmost to make it happen!

4

"IF WE KNEW WHAT IT WAS WE WERE DOING, IT WOULD NOT BE CALLED RESEARCH ..."

ALBERT EINSTEIN

Back to Basics

I had not physically worked in a laboratory for a good number of years and so it was with some trepidation that I donned a white lab coat in Ted's laboratory at the University Park Campus and began to relearn laboratory techniques I had not used in 20 years. The only reassuring thing about it was that the basics of microbiology, the aseptic technique, plating, counting had not fundamentally changed either and it was with some joy that I got stuck in. As my coach, I had Phil Stone and as mentor Ted Cocking and I felt that I relatively quickly got back to at least a competent level of practice. This was in itself a personal achievement but also for the research we needed to undertake, an absolute necessity because the first field season was no more than 3 months away. I was under immense pressure to deliver a formulation of the Gd for application as a seed inoculant for the field season that would start in March 2013. Given it was now the start of November 2012, this did not give me much time to get to grips with the laboratory practicalities, designing and implementing a programme of research to confirm the principles established by Ted and then to design, develop and test a formulation.

The first decision I needed to make related to the choice of plant species to use in our bioassays. We needed to establish a model system to evaluate how well Gd colonised and impacted on plant growth. At that time I was concerned that the bacteria would possibly take too long to colonise throughout a rapidly growing annual crop plant within a single season. Certainly it would need to colonise, establish and distribute itself sufficiently around a plant for it to have an influence on crop yield. Although in fairness, Ted and Phil had already demonstrated in a pilot field microplot experiment with maize that yield differences were possible within a season (Figure 4.1). However, given that time is always precious in commercial R&D and I could not afford to lose a season to a failed trial, I decided to use a perennial ryegrass as the test plant. The reasoning behind this was three fold. Firstly I had developed some familiarity with this grass species while studying for my PhD on the grass aphid *Metopolophium festucae cerealium* and more importantly, because *Lolium perenne* is a perennial species we could, if necessary, run a single trial over a number of seasons and the yield assessments (measured as the biomass of cut grass) could be made at multiple points during a season, rather than waiting for a single measure of grain yield at the end of a season. For this same reason grass made a good laboratory plant because we could assess the effect of Gd colonisation on plant growth using a measure that would directly relate to field value – namely the amount of foliage produced per plant. Time was of the essence so working with grass where a high turnover of relevant measurements could be made in a relatively short space of time was invaluable.

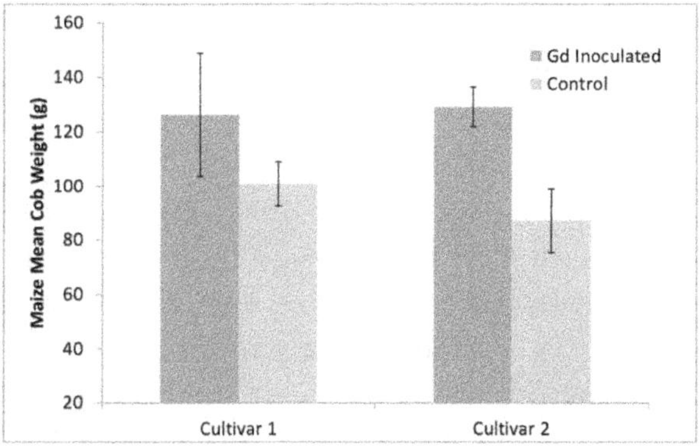

Figure 4.1. The first micro-plot unreplicated field assessment of an inoculation of Gd on seeds of two varieties of maize, carried in the UK. The pilot suggests that cob dry weight (means of 10 plants) of maize is higher in the inoculated compared to the control plants (Reproduced by kind permission of Ted Cocking).

In order to measure the impact of Gd nitrogen fixation through the indirect measure of enhanced biomass in the field we first needed to establish that inoculating seed with the bacteria did in fact influence the subsequent vegetative growth of grass in the laboratory. To this end Phil and I set up a number of experiments where we added all the normal nutrients required for plant growth, except nitrogen, in a sterile vermiculite and sand mixture. This enabled us to evaluate the difference in grass growth when seeds had been inoculated but had no access to sources of nitrogen other than from the air. By simply inoculating seeds with the bacteria and then growing up these plants and comparing their leaf dry weights at the third leaf stage, we were able to determine if Gd affected leaf growth when the plant had no access to nitrogen other than that in the seed or that from the air, which only the colonising bacteria could access.

What we found was very encouraging and the results in Figure

4.2 show that the seeds inoculated with the bacteria produced plants with a significantly higher above ground biomass relative to the controls. Although a positive result, we could not at this stage be sure that the Gd was actually fixing nitrogen, the biomass differences could just as easily be explained by the release by the bacteria of plant growth hormones – which Gd is known to do. Irrespective of this, the difference between inoculated and control plant biomass gave us an indirect measure of the impact of the Gd - assuming we were achieving intracellular colonisation of the inoculated plant. This we were later able to demonstrate by the use of a so-called GUS-labelled strain of the Gd that shows up under the microscope with the appropriate colour stains.

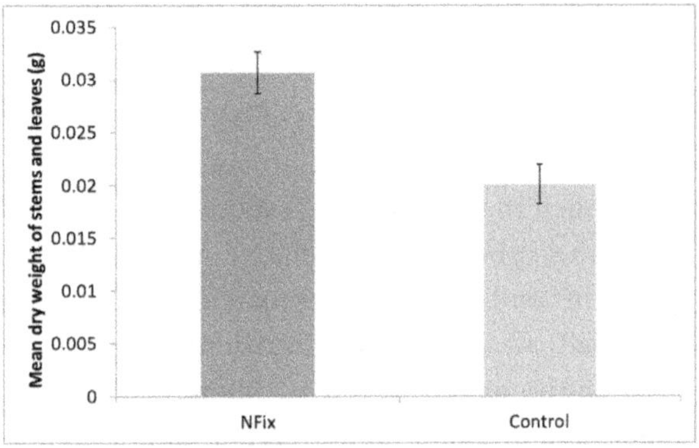

Figure 4.2. Dry weight of third leaf stage *Lolium perenne* seedlings grown in a reduced nitrogen media (vermiculite and sand mix) to determine any impact of inoculation with the bacteria Gd. The 35% difference in the amount of foliage of the NFix treated plants is significantly different at P<0.001. The vertical bars represent the variation around the mean as a standard error (Dent, 2013).

Although an initial positive result, it was still a long way from having a formulation that would function under real-life situations in the field where other sources of fixed nitrogen would be available to the plant in the soil. I needed to come up with a formulation that we could use to treat the seeds and use in field trials that were now only a few weeks away.

With some knowledge of microbial formulations from my days managing the locust LUBILOSA biopesticide development programme based on a fungus *Metarhizium anisopliae*, I knew we needed a number of essential ingredients to make a product function. This would generally include a sticking agent to hold the bacteria onto our target surface (in this case a seed) and a surfactant that would help ensure sufficient spread of the bacteria over the seed so that we did not get clumping of the bacteria. We were placing such low numbers of bacteria on each to seed the importance of a surfactant could not be underestimated. I had previous experience with a surfactant called Polysorbate 80 with the locust programme so I opted for this as a starting point. A commonly used sticker that is known to be compatible with gram-negative bacteria is a natural substance called gum Arabic, so this again seemed like a good option. We also were aware that the University of Nottingham patent used 3% sucrose which acted as a carbohydrate energy source for the Gd, so 3% seemed like a sensible starting point for sucrose. However, as to the amounts of the Tween 80 and gum Arabic that should be added - well a wide range of concentrations was technically possible, but given the time constraints, Phil and I did not have the luxury to range test different amounts or to determine compatibility of the three ingredients. It really was a 'stick my finger in the air' moment and it was necessary for me to make an educated guess as to what might work. And this I did; the technology that became known as NFix[1] was thus created, with a little bit of science and a little bit of magic - the latter otherwise known as guesswork!

First indications for the NFix formulation were positive (Figure

4.3). I was pleased with this as a first step and was able to present the initial results later in the year to the Ontario Agri-Food conference in Guelph, Canada entitled "Nitrogen fixation in corn and 199 other crops" in April 7th 2013. What was interesting about the results from the experiment depicted in Figure 4.3 was that even in the presence of unlimited nitrogen in the media, the dry weight of the above ground foliage increased relative to the uninoculated controls by as much as 35%. After this initial successful result I had little time to perfect the model system or to conduct many further assessments - the formulation of sucrose, gum Arabic, Tween 80 and Gd was what we would use to test the NFix technology under field conditions.

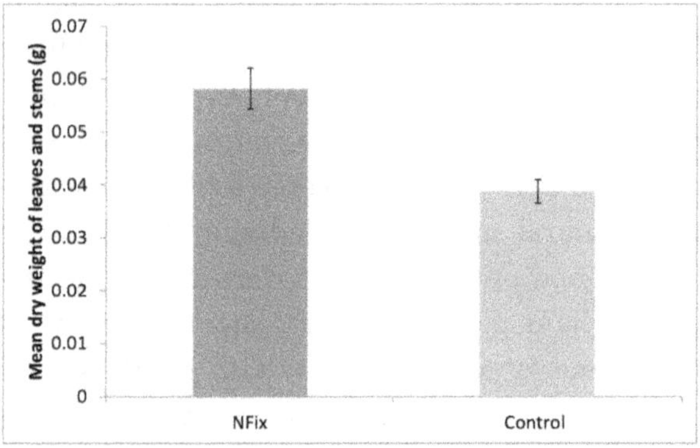

Figure 4.3. Dry weight of third leaf stage *Lolium perenne* seedlings grown in normal compost with (with excess nitrogen) to determine the impact of NFix treated seed compared to untreated controls. The 34% difference in the amount of foliage of the NFix treated plants is significantly different at $P<0.001$. The vertical bars represent the variation around the mean as a standard error (Dent, 2013).

First Field Trials

I have over the years written and published on the subject of the great gulf between successful controlled laboratory experiments and the variable results that often accrue from field-based research. My first foray into this subject was a paper I presented to a conference in 1990 of the Association of Applied Biologists entitled "*Bacillus thuringiensis* for the control of *Heliothis armigera*: Bridging the gap between the laboratory and the field." (Dent, 1990). At that time and every time since then when I have waxed lyrical about this subject, it is to argue that it is very difficult to predict from laboratory experiments under tightly controlled conditions, what will happen in the more real-world situation of a properly conducted field trial. Also, and I made clear in two presentations on the topic in 2017 Industry Conferences on the Microbiome in Raleigh North Carolina and London UK (Dent, 2017b & c), the laboratory tends to focus on simple single indicators of efficacy measured in an experimental context which (i) may not elicit or reflect the expression of the same function in a 'natural'/cropped system or (ii) focus on single indicators of effectiveness that do not reflect the multitude of interacting factors that may play a part in the field situation or those that address commercial need. And because of all this it is often the case that excellent results in the laboratory fail to translate into anything but mediocre results in the field. However, this is not always the case - sometimes - just sometimes, the converse can be true but would this be true with NFix; would the results we had obtained with the NFix formulation in the laboratory hold up in the more real world field situation? I had no idea, just hopes!

Our first Azotic field trials with NFix were carried out with the assistance of the University of Nottingham field trials team John Alcock and Matt Tovey. The trials were set up as nitrogen equivalence trials which means the impact on the crop of the NFix inoculated seed would be compared against uninoculated seed (the control) and uninoculated seed with different levels of nitrogen fertiliser applied as a percentage (25, 50 and 75%) of the nitrogen fertiliser recommended rate. The trial was set up in this way with our limited resources to give us an indication of how much nitrogen fixation by Gd could substitute for nitrogen fertiliser use. At this point we needed to know if the nitrogen fixation by the bacteria would be equivalent to more or less the same as half the normal amount of fertiliser. We had no idea if this would be the case and so this first field experiment would give us an indication of just how good Gd might be at influencing crop growth and yield.

More fundamentally, this trial was the first use of NFix in a replicated field trial and to be honest I was just hoping that there was colonisation of Gd and some level of measurable impact. We were working with three crops, grass (*Lolium perenne*), wheat and oilseed rape (OSR). We had not managed in the 3 months that we had for laboratory research to carry out any experiments with the bacteria or NFix with wheat or OSR, so we were essentially running blind on these crops. It was going to be a highly fraught summer field season!

Oh for Sunny Days

UK summers had been generally dull and wet for six consecutive years since 2007 and so the prospect of weekly field sampling for 5 months from March 2013 did not present an enticing prospect for me. I was travelling to Nottingham every week for a couple of days in order to work with Phil in the field to collect all the data we needed from the 3 crops we had treated. I thought I had left such intense fieldwork behind many years ago in my youth, and I know Phil felt the same way. However, it became a routine we came to enjoy, not least because as the weeks progressed the sun continued to shine. Providence must have been on our side because it turned out to be the sunniest summer in the UK since 2006, with sunshine totals at 114% of the long-term average. I even achieved an impressive suntan!

For those who love technical detail or just would like to visualise the trials, each of the wheat and OSR plots were 12m x 3m, while the grass plots were each 6m x 1.6m; there were 6 treatments and four replicates of each treatment that included zero N (residual soil N levels only) and full N which was compared with NFix inoculated at zero N, 25%, 50% and 75% of the recommended rates of N. The question that was being asked was what level of nitrogen plus NFix provided yields equivalent or higher than the recommended rate of 100%N? Soil nitrogen tests had been taken across the field sites and rates of fertiliser determined from the residual soil nitrogen based on the recommended rates for each crop.

Measurements were made of growth stage, plant height, leaf greenness, number of tillers, 1000-grain weight, plot yield, for wheat and OSR. For the grass plots the measurements were sward greenness, biomass per plot (taken every two weeks using a mower) and leaf nitrogen content.

Phil and I took the measurements and samples of grass at pre-determined intervals (with some variation due to the weather - one cannot mow grass in the rain or when the sward is too wet) and collated the data; each week trying to see what might be the outcome and trends of the experiment. As anyone who under-takes crop trials knows the summer sampling period can be a tiring time. Phil and I were frequently exhausted by long days of sampling but as the results started to accumulate we became more and more encouraged and even a bit excited.

As part of the measures we had also invited a postgraduate student Gary Devine, who was particularly adept at PCR analysis a technique that would identify Gd DNA within the plant material. Gary took some grass samples from the field in order to determine whether any of the leaves had been colonised by the bacterium. Thankfully Gary was able to find Gd in these leaf samples and this gave us some confidence in our results that were demonstrating increasingly remarkable benefits.

Results beyond expectations

I would have been satisfied with our first season of field trial results to just be able to demonstrate that if we coated seeds, sowed them under field conditions and that we were then able to identify the presence of the bacteria in the growing plants. That we achieved something beyond this expectation was a genuine surprise. The results from grass were the most impressive with an increase in total biomass produced by nearly 50%, and increase of leaf nitrogen content by over 40% and greenness (measured as a normalised difference vegetative index known as NDVI) increases of 10% (Figure 4.4).

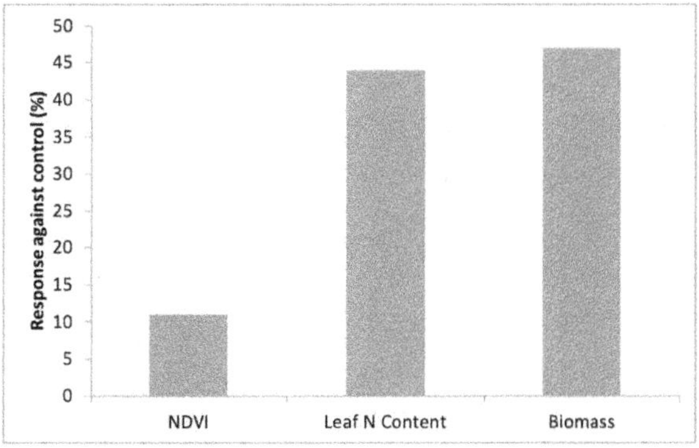

Figure 4.4. The percentage increase in NDVI, Leaf Nitrogen Content and Biomass a grass (*Lolium perenne*) treated with NFix only compared to the recommended rate of nitrogen fertiliser (Dent 2015, 2016, 2017a).

The results from wheat and OSR were no less dramatic. The yield differences between 100% nitrogen and 50% nitrogen plus NFix were not significantly different demonstrating that the benefits of the Gd bacteria were equivalent to half the nitrogen fertiliser applied (Figure 4.5).

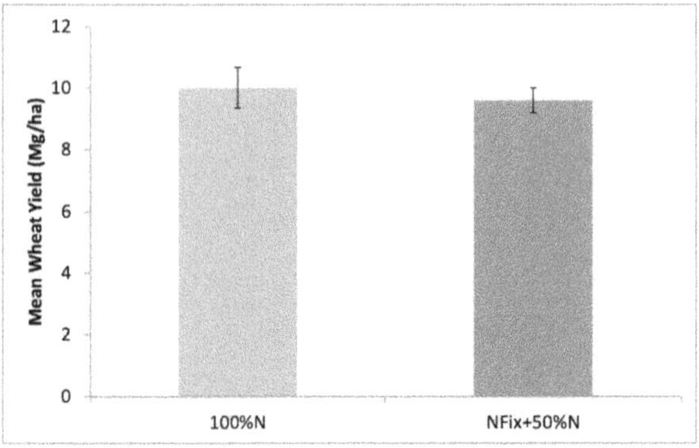

Figure 4.5. Spring sown wheat yield in 2013 comparing NFix combined with 50% N of the recommended rate of nitrogen (100%). There was no significant difference between the two treatments which demonstrates the nitrogen fixation by the Gd is equivalent to around 50% of the recommended rate of nitrogen fertiliser. The vertical bars represent the standard error of the means (Dent 2015, 2016).

"There's no such thing as bad publicity"

We had achieved much more than anticipated from our first year of field trials. Having hoped to just be able to demonstrate NFix treated seed sown under field conditions could facilitate colonisation by the Gd in the growing plant in one crop, let alone three, would have been an achievement, but to have been able to demonstrate an impact on yield was a major bonus. At such an early stage in the product development process, it really did bode well for future developments. However, in order for there to be any further developments, Azotic Technologies required another round of funding. To achieve this we needed to raise the profile of the Company and awareness of the potential of the technology among prospective investors - and a decision was made to take a high-profile publicity route to achieve this.

Scientists, as a generalisation, are often more than a little cautious about use of the media for presentation of results, not least because although it may be the marketing guys who set up and ensure adequate publicity, every CTO gets to answer all the technical questions. Given that scientists tend not to want to talk about certainties but rather probabilities and the latter are the last thing that the media like to deal in, we essentially trade in different currencies. Thus it was with some trepidation when it was agreed with the University that a press release should be issued about NFix, Azotic Technologies and our field trial results, I knew there would be fall out - but it was going to have to be a necessary evil if we wanted to secure another round of funding.

And there was fallout!

The University of Nottingham released their statement on July 25th 2013 entitled "World changing technology may enable crops to take nitrogen from the air" making claims about *Gluconaceto-bacter diazotrophicus*, its intracellularity, nitrogen fixation and its potential to substitute for nitrogen fertiliser use in any crop. The release received really good coverage by the media, perhaps as one would expect with such important claims. But not all of the responses were positive!

One of the headlines associated with the negative fallout stated "Worst science by press release of the year: nitrogen fixation". There were genuine concerns about the apparent lack of evidence - which in fairness we were still waiting for from our field trial results but we were getting positive responses by that time and were feeling pretty confident. In hindsight it might have been too early and some of the criticisms were perhaps justified.

This media attention created a rather public problem for me as I joined a queue for the Cholmondey Room for an event at the House of Lords. A scientist with well-known and espoused opinions about the nitrogen fixation research of Ted Cocking decided to voice these as I stood in line behind him waiting to pick up a glass of something bubbly. The individual proceeded to lay into me in a rather loud voice that brought our conversation to everyone's attention. The individual, who will remain nameless, stated that we had presented no data that he had seen (as though he was the sole arbiter of scrutiny) that supported any of our claims. I am well used to the arrogance of some scientists but it has to be said the vehemence with which this diatribe was delivered was a surprise even to me.

Typically for a scientist, the individual was helpfully able to list all of the experiments we needed to carry out that would convince him we had something of import. Crucial amongst these were the GMO *nif*-minus gene and labelled nitrogen studies to demonstrate whether if the nitrogen-fixing genes were switched off, the plant growth was still as dramatic.

I knew the scientist well enough, and I remain respectful of his achievements, but it was clear during this monologue that this feeling was not in any way mutual. I realised then, as now, that this individual's passion was in part driven by his need to defend a dying paradigm. It was also evident that if what we claimed was true, then it would reduce the need for his own research and hence, there was an element that was motivated by vested self-interest. The strident views he offered were not necessarily just about scientific objectivity. In fairness to him, and I had little reason to be fair, we knew we had data yet to collect, but something positive came from this interaction that helped me to identify the key research outputs required to demonstrate intracellular nitrogen fixation in non-legumes with Gd in order to satisfy even our most strident critics. On the upside, the publicity that had so offended my scientist colleague had also increased our profile and we were making headway in discussions with investors.

Karel and Koppert

Early in my career I had trained as an entomologist. My PhD thesis was on the biology of a grass aphid, and as a Postdoctoral internship at ICRISAT, India I worked on pheromone monitoring systems, and in the late 1980s I had taken it upon myself to write a text-book on *Insect Pest Management*. I was 28 and unemployed for a year before I started as a Research Associate at University College Cardiff during which time I undertook this ambitious exercise and it occupied every moment of my spare time for 5 years before my *magnum opus* was published in 1991 (Dent, 1991). During that long and laborious process I was privileged to get to know some aspect of every discipline involved in insect pest management. And because I had reviewed the whole subject it provided me with a holistic perspective that it is often difficult to achieve as a young early-career scientist. One of the subjects with which I became familiar was the subject of biological control, both from a theoretical and practical point of view. Although I did not realise it at the time, it was one aspect of insect pest management that would come to influence the rest of my scientific career.

In getting to grips with the scientific literature on biological control for Chapter Six of the book, there was one company's name that kept recurring, Koppert Biological Control Systems. Koppert BCS (Koppert for short) was a small Dutch company based near Rotherdam in the Netherlands. What made Koppert so special was that the insect pest control sector during the 1960-70s was dominated by nine or 10 multinational agrochemical companies, all of whom developed and sold chemical insecticides whereas Koppert was selling living-insects referred to as natural enemies. Initially they focused their business on a small parasitoid for control of tomato whitefly and a predatory mite for control of the spotted spider mite, a pest in cucumbers and other horticultural crops, largely in glasshouses. This may not seem notable except when one realises that Koppert's products were incompatible with the use of insecticides and hence, they were operating in competition with the large agrochemical companies. While the theory and technical knowhow supporting and underpinning use of natural enemies of pest insects was well known there were few industrial pioneers in the sector. This marked Koppert out as something exceptional - a company and individuals prepared to challenge the paradigm, and they had apparently done so rather successfully!

Back to 2013, the publicity surrounding our field trials came to the attention of Dr Karel Bolckmans who was at that time in charge of product development at Koppert. Karel, like me, immediately grasped the significance and importance of a technology that could potentially replace nitrogen fertilisers. Karel is not a shy retiring individual, and he had no qualms about getting on the phone to Peter Blezard to arrange a meeting. This came completely out of the blue for us, but as Peter rarely refused a meeting one was arranged in London for late November.

So it was that on a cold Friday November morning, Peter and I waited in the Institute of Directors (IoD) on Pall Mall, London for our meeting. The flight from Amsterdam was delayed by fog and it was not until lunchtime that Karel Bolckmans and the Director of Koppert, Paul Koppert walked into the grand setting of the

Directors Room of the IoD. Here, in Paul Koppert was the Director of a company that had for over 40 years challenged the agrochemical industry and their business model and, not only survived the experience, but also carved out a whole new sector of business based on shipping living insects and microbes all over the world. This was someone whose family was, in my mind, a legend in its lifetime. And now they were interested in our own pioneering work on nitrogen fixation. Paul and Karel were completely different characters. Karel was flamboyant, effervescent, large and larger than life, while Paul more diminutive in stature but not in statesmanship, gentlemanly and unassuming. Somehow this felt right from the start, a completing of a circle for me, a linking of my early career with this point in time 30 years later at its other end.

Suffice to say this meeting and every subsequent meeting went well and to keep the story moving, an investment was agreed, signed and sealed in January 2014. This significant event was the game changer needed to really accelerate development of Azotic and our nitrogen-fixing technology. Koppert had both formulation and manufacturing capability and was an ideal partner for Azotic.

---

1. *N-Fix® is a Registered Trade Mark of Azotic Technologies Ltd.*

5

WHO OWNS' WHAT, IS IT SAFE AND CAN WE USE IT?

Another day, another email, another possible piece of information that finds its way on to my desk, the finding that could slow our journey, send us on a nightmare detour or derail us altogether. Is it a company releasing a competitive product, or publication of the piece of research that counters everything we claim, the flaw in our logic? I never knew what was coming, what questions I would have to answer what insult to my intelligence I might have to ignore to take us one further step forward each day. I enjoyed every moment of it in the way a boxer keeps entering the ring because he believes he can win and will keep picking himself up off the floor while there is a chance this may be so. Sometimes being a CTO in a technology start-up it is a bit like being 'punch drunk', one keeps getting up although each time one feels a bit more dazed. There is always the belief that the opponents will wear themselves out and the head will clear eventually!

Any science-based innovative company is not just about the technology it is developing but rather it is also about the secondary issues that arise because of the new development. The greater the perception of difference between what is thought to be known and what is thought might-not be known, the greater the intensity of interest in the potential size and nature of that difference. If a disruptive technology is under development then that difference may be perceived to be very large indeed.

At the beginning, while everything is exciting, it is easy to be dismissive of the secondary factors that may play a role in the delivery of a technology to market. Assumptions are made and provided there is some logic and evidence to support such assumptions they tend to be left well alone on the periphery of things. As interest grows and the prospect of external support increases, the assumptions tend to come under closer and closer scrutiny.

The questions that are asked are not just about whether the technology may or may not work, but in addition – does it belong to you, is it safe and can it be used? For technologies such as NFix that are based on a living bacteria then there are lots of questions that can be asked. Given that the bacteria was originally isolated from sugarcane in Mexico; is limited in its host range and geographical distribution; has to be manufactured and distributed prior to its release into the environment; when it is applied to crops, and those crops or products derived from them, will be consumed by humans and/or animals – then regulators, manufacturers, distributors, food processors and consumers all have questions that they want answered. So does the Company have rights to commercially exploit the Gd, is it safe to consume and ultimately is it ok to release this microorganism into the environment. Failing to have answers to any one of these factors could alone stop NFix in its tracks irrespective of how good it may be, or the scale of benefits it may bring. NFix was in the firing line not just because of the underpinning science or the disruptive nature of the technology but because of a whole host of other secondary but important factors. It was incumbent on me as CTO to have all the answers!

Freedom to Operate

When in discussion with potential investors and partners, one of the key queries that invariably arises is associated with the concept of a 'freedom to operate' (FTO), and we were no different in Azotic. The FTO concept is relatively simple, referring to any action, such as product evaluation, that can be carried out without infringing any intellectual property rights of a third party. For example, if a competitor company owned a patent that involved use of Gd or another bacteria in a specified and particular way and Azotic used that method to facilitate commercial use of a product, then that would be considered a patent infringement. Knowledge of the potential for such IP infringement is important so that either alternative methods can be used or a license for use of the IP can be negotiated. Either way, this would have an impact on the Company. Usually in an SME (and it was certainly the case at Azotic), the only person with sufficient technical expertise to make such judgements or to assist a patent attorney in evaluating FTO will be the CTO. I spent many hours poring over the details of each patent that we licensed, getting to know, each of their limitations or potential areas for dispute, identifying appropriate responses to highly detailed technical matters and then as many hours again in very precise line by line assessments with investors' IP lawyers. The importance of FTO cannot be underestimated because it forms the bedrock of the Company's IP strategy and the priorities given to specific areas of R&D. Each new patent we found and each new research finding that emerged from the literature or from enquiry did not invalidate our approach or our claims. So with growing confidence, we began to realise we had clear blue water between us and everything else that had been or was being developed. Although, even with this knowledge, I did not start to rest easy for two or more years, when I began to feel that nothing untoward was likely to jump out at us.

Convention on Biological Diversity and Benefit Sharing

If you do not have the right to use a microorganism because it belongs to someone else then it cannot be commercialised. To do so has serious ramifications, not least a failure to comply with international conventions. As it happens, I am fairly well informed about this area because of my former role as a Director and then Managing Director of CABI Bioscience. This amazing intergovernmental, international organisation manages a microbial culture collection where it maintains third-party patent deposits of microorganisms meeting all the pertinent international treaty standards and obligations. Prime amongst these, is the Convention on Biological Diversity (CBD) and the associated Nagoya Protocols on access and benefit sharing.

During my time at CABI Bioscience the curator of the collection was Dr David Smith who is a world-renowned expert and campaigner in the field and it was he, in his own inimical way, who imbued in me an absolute and unwavering commitment to the CBD and its proper and effective implementation. So it was that when I founded Azotic I needed to be assured that we were fully compliant with the CBD in our dealings with the use of Gd the strain of which had originated in Mexico. If we were not, or we could not find a way to proceed, then the project was a non-starter.

Strains of micro-organisms from all over the world are exchanged between academic researchers in order to facilitate research. However, these strains are gifted on the understanding that the knowledge produced is for public dissemination for the greater good. If IP arises that could be commercially useful, then strains are usually supplied under a material transfer agreement (an MTA) where the terms of use and exploitation are defined. These processes are gradually being adopted globally but back when Ted first collaborated with Dr Jose de Jesus Caballero-Mellado of the De Centro de Investigación sobre Fijación de Nitrógeno (CIFN), Mexico (which changed its name to Center for Genomic Sciences (CCG) at UNAM in 2004), there was not a thought by either party that anything other than useful research would result. Ted took receipt in October 1998 of a number of Gd strains from Jesus that he had isolated from sugarcane in Mexico. It was one of these, some years later that was to prove vital to the recognition of intracellular colonisation and its implications. Sadly as time moved on so Jesus passed away in 2010 prior to Azotic arriving on the scene, and so he never got to see or to know how his initial collection of strains from sugarcane in Mexico became set to change global agriculture.

Given my background with CABI Bioscience I wanted to formally ensure Azotic was wholly committed to the CBD and the sharing of benefits arising from commercial exploitation of biodiversity. As demonstration of that commitment, on the 20th December 2012 the Azotic Board unanimously agreed to prioritise distribution and sale of the NFix technology to Mexico and to establish a Fund for the furtherance of biofertiliser based research capability within Mexico.

I have since travelled to Mexico and visited Ministry officials with Nick Gosman in order to pursue collaboration with the Mexican government, philanthropists and the International Maize and Wheat Improvement Centre (known by its Spanish acronym CIMMYT) on projects for the introduction and use of NFix in Mexico.

From the University of Nottingham point of view, they too had a CBD and Benefit Sharing Policy. As part of this Alan took charge of negotiating an inter-University benefit-sharing agreement with UNAM. The details of benefit sharing is confidential between the two Universities but UNAM will receive a revenue share as a proportion of the revenue generated from payments based on the licensing agreement between the University of Nottingham and Azotic Technologies Ltd. As much as is practically possible, everyone involved, benefits.

So it is that I ensured, as much as possible that Azotic is compliant with the international conventions on biological diversity and benefit sharing. It was therefore, both frustrating and annoying that I started to pick-up from comments being made at conferences that suggested we were not compliant with the necessary conventions. Criticism of Azotic by detractors was targeted at the idea that the strain of Gd used by the Company had been obtained from Brazil. Anyone conversant with the issues surrounding the CBD and benefit sharing are aware that Brazil is very sensitive about protecting its immense biodiversity, not least that from the Amazon rainforest. The Federal Government quite rightly takes great care to restrict access and exploitation of its biodiversity without formal agreements being in place. The insinuation of the whispering campaign was that Azotic had somehow obtained our Gd strain by dubious means and therefore had no rights to its commercial exploitation. Rumour-mills can be quite divisive and so it was therefore with particular delight, that following a presentation given at a workshop at Rothamsted Research in the UK, that I had an opportunity to lay the matter to rest. In a rather triumphalist tone a scientist stood and asked whether our Brazilian strain of Gd had been obtained in accordance with the conventions of the CBD? It is not often one gets the opportunity to confront the issues of ones critics head-on, but this was one of those rare occasions. I just hope that in explaining our obligations, commitments and compliance my response was not too condescending, but perhaps, just once I may be forgiven if it was!

Generally - Is it Safe?

No one would deliberately want to develop a product that is not as safe as it can be, but that does not mean that any product will be safe under all conditions, situations and uses. There will always be attendant risks in any product - it's just a question of what those risks are and the extent of the hazard they pose and any trade-off there may be from the level of benefits that accrue from their proper use. My interest is in developing and using a safe product for both people (and animals) and the environment. I would not have engaged in the project without a strong level of confidence in this. However, demonstrating that the confidence one has is justified and evidence based is important in product development, so that any missing areas of concern can be identified and addressed.

In working with microorganisms one is always sensitive to, and aware of, potential threats to a working environment. One of the things that helps us assess risks with working with such microorganism, and the first step for any scientist, is the laboratory safety rating provided for each particular class of organism. This was my first 'port of call' with regard to Gd, and I was not surprised to find that Gd comes into the lowest category of hazard for use of bacteria in a laboratory. Species of *Gluconacetobacter* are classified according to the national and international Culture Collections (e.g. the American Type Culture Collection ATCC) as being Biosafety Category 1. This category of bacteria means it is well-characterised, not known to cause disease in healthy adults, and is of minimal potential hazard to staff and the environment. The advantage of working with a bacterium in this safety category is that the level of precautions that have to be applied are the lowest for a microbiological laboratory. In reality this means that the laboratory does not even need to be separated from the general traffic patterns in the building, work can be conducted at the bench using standard microbiological practices and any contaminated materials can be placed in open waste receptacles.

Other useful criteria that can be used to assess the general safety of a micro-organisms are published standards. For example, the USA Food and Drug Administration has a system of regulation for substances derived from microorganisms that are *Generally Recognised as Safe* (GRAS): *21 CFR184* includes acetic acid produced by fermentation. The most explicitly known and most widely applied industrial strain of acetic acid bacteria belong to the species of *Gluconacetobacter* (Stasiak & Blazejak, 2009; Schuller *et al.*, 2000). Hence, species of bacteria in the genus *Gluconacetobacter* that includes *G. diazotophicus* (Gd) are covered by the FDA GRAS regulations. There are additional examples such as the use of species to produce cleaning products (Attwood *et al.*, 1991; Stephan *et al.*, 1991) and production of an extrapolysaccharide fructan that is used extensively in the food and pet food industry.

Can it be Consumed?

It is never sensible to be complacent about food safety, and not wanting to sound facetious, if anyone asked me if have I consumed Gd, I can honestly say that the answer is "yes". In fact, it is not difficult to find oneself in a situation where it is possible to consume this particular bacterium. This is because Gd is found quite naturally in sugarcane and a range of other high sugar content fruit and some vegetables. For instance, Gd has been found to naturally occur in pineapple, sweet potato (Paula *et al.*, 1991), mango fruit (Rocafull *et al.*, 2016), beetroot (Muthukumarasamy *et al.*, 2002), carrot and radish (Madhaiyan *et al.*, 2004), and while one might eat these fruit and vegetable in a cooked or processed state, if they have been consumed in their raw form, then live Gd will have been consumed. And this has probably been the case for millennia.

With particular regard to sugarcane, anyone who has sucked, chewed or drunk the raw sugar juice from a cut stem will have consumed Gd. The raw juice is extracted and sold by street vendors and consumed (without pasteurisation) in many countries in Southeast Asia, South Asia, Latin America, Brazil and East Africa. Interestingly the raw juice is considered as having health promoting qualities, with benefits that include mitigating against cancer, stabilizing blood sugar levels, assisting in weight loss, reducing fevers, preventing tooth decay, and a host of other health benefits (McCaffrey, 2011). One last example of the consumption of *Gluconacetobacter spp.* is that of its role producing through fermentation the drink kombucha tea which is not pasteurised. The USA FDA and CDC have investigated all reports of health concerns with regard to kombucha tea and consider it to be safe to drink.

Can it be Released?

Gd has been isolated from a variety of crops around the world, with its distribution most common in tropical and subtropical environments in high sugar content crops. This means that Gd will have a limited distribution and hence, from an environmental perspective it is important to understand what level risk may be posed by the introduction and release of this bacteria in countries where it is not indigenous. The concern with introducing new or alien species (which is how they are referred), is that in their new environment without the biotic and abiotic limitations of their natural environment, they may become invasive and thereby pose a threat and cause harm to the indigenous organisms and the environment. To knowingly introduce such an alien invasive species is unlawful, but that aside, any biologist would find it reprehensible to facilitate such a calamity. Not least a biologist with my own expertise and background. The control of alien invasive species by biological control methods was one of the other main activities of CABI Bioscience (I mentioned above), of which I was a former Managing Director. CABI Bioscience has made many successful biological control introductions to re-associate alien invasive species with indigenous natural enemies to facilitate their management in their alien environment. Knowing what I know, and with my experience, I investigated in some depth to convince myself that Gd could not become an alien invasive species and cause harm wherever it was to be used. To ensure this research could be made available to government agencies with an interest, I prepared a report of my findings -

which were thankfully supportive of use of Gd in alien environments.

The gist of my assessment is based on what is known about the bacteria, its habitat, its survival and transmissibility all linked to how we proposed to apply it. First of all, the general consensus from collating the work of many scientists is that Gd in an obligate endophyte, meaning it only thrives well within its host plant and not outside of it and evidence certainly demonstrates that the bacteria is rarely found in studies of soil and weeds around sugarcane. The ability of Gd to therefore reproduce in the explosive way typical of alien invasive species in, say, the soil, is just not tenable. If it cannot then disseminate widely, then its invasive potential is definitely extremely low. Given that Gd has been moved around the world in cuttings of sugarcane and reproduced in new environments by vegetative reproduction the bacteria has had ample opportunity to display invasive qualities moving from plant to plant has failed to do so. Gd seems to have a relatively narrow host range and perhaps part of the reason for this is that, as we found, it is not easy for the bacteria to colonise species and it requires the help of the NFix formulation to achieve this. Since the NFix is applied to the seed prior to planting and the bacteria have poor survival on the seed, this reduces the risk of survival in the soil environment, so use of the product will not dramatically increase the risk of between plant transmission.

Interestingly, it is quite difficult to achieve inoculation of the roots with the bacteria, which is why the NFix formulation is so important in helping to achieve this, so it is unlikely although not impossible that the bacteria will readily colonise different plant species from the soil. Transmission may be possible through insect feeding, and the attributes of the bacteria suggests it might survive in insect guts and this is an area that will in due course require further research and consideration. However, never wanting to take anything for granted, and in spite of the small risks involved I made sure that research programme investigated these risks further so that information is at least available should regulators show interest. Hence, the risk to the environment and biodiversity is minimal for this non-invasive obligate, diazotoph. This is especially the case when placed alongside the potential environmental benefits of its use in the reduction of nitrogen fertiliser use and subsequent reductions in nitrous oxide emissions and nitrate pollution: all of which represent major threats for our climate, biodiversity, waterways and drinking water.

And Finally...

Overall, I was confident that with the information we had available that the likelihood of infringing others' IP was low; that every attempt has been made and meaningful agreements and commitments had been set in place to ensure that benefits of indigenous biodiversity are to be appropriately shared; and that the technology is as safe to use and to consume as can be. Additional research in this area should continue however, in order to further substantiate and validate this view. The importance of this cannot be underestimated.

6

"DECIDE WHAT YOU WANT, DECIDE WHAT YOU ARE WILLING TO EXCHANGE FOR IT. ESTABLISH YOUR PRIORITIES AND GO TO WORK."

H.L. HUNT

"Wherever I lay my hat ..."

Nottingham is a great city. Aside from the obvious and popular myth of Robin Hood and the 'Sheriff of Nottingham' that immediately springs to mind, the city has a genuinely rich history and culture. It is a living, thriving city that is proud of its heritage not least that associated with Jesse Boot the retailer bearing his family's name, and its two universities 'Nottingham' and 'Nottingham Trent', the former established on land bequeathed by the Boot's family. As I mentioned earlier, Ted Cocking had been a member of staff at the University of Nottingham since 1959. The present day institution has moved with the times and can boast an unrivalled strength and depth of expertise in the agricultural and related sciences, exemplified by its facilities and capabilities at its campus at Sutton Bonington just a few miles from the city.

When, with sufficient investment we were able to envisage establishing our Company laboratories, I was given the choice as to where in the country to base our in-house capability. Living in Fleet, in Hampshire, it might have been tempting to have suggested a laboratory somewhere relatively close to home, the University at Reading maybe, or Surrey University at Guildford, both within half an hour driving time of my home. But instead I realised that the potential benefits of being able to work close to home were nothing compared to the benefits of basing ourselves in Nottingham. It is a decision I have never regretted even though 6 years on, I have on average spent at least half of every week living away from home.

The choice of location for our laboratories was in the end, relatively simple. A University town was required with the relevant scientific discipline to provide a potential source of qualified scientists, technical knowhow and potential academic collaborators and given Nottingham is one of or the leading agricultural science Universities in the UK, it was an obvious location for us. The nearness of Ted Cocking and his laboratory and the ability to readily access Ted's expertise and knowhow was absolutely essential. It was also my belief right from the beginning that the NFix technology could truly revolutionise global agriculture, and if we were to make that happen then it would only be right and proper that that achievement was wholly associated with the University and the City in which the initial discovery and proof of concept had been achieved. I am a great believer in the role of legacy in life, and because of this I wanted to ensure that the link with Ted, the University and the City of Nottingham was preserved in all we achieved. And during my time at Azotic that was always a prime driver of mine in promoting all three as part of our Company's mission.

Most Universities these days have an associated Research Park in order to generate a hub that includes academics, spin-out companies and where industry start-ups can build their business with access to a wide range of resources. Nottingham was no different except one of the city's resources was an incubator facility near its centre on Pennyfoot Street, known as BioCity. Here was another legacy of Boot's, since BioCity had originally been the R&D facility for Boots and indeed the building hosts a blue plaque at its entrance to recognise the discovery of the pain killing drug ibuprofen, on the very premises. I had visited BioCity some years earlier to meet with the Director Glen Crocker MBE to discuss UK innovation systems, and I held the man and his achievements at BioCity in very high regard. It was an easy decision, with the facilities on offer to agree to establish Azotic Technologies R&D capability at BioCity and on the 3rd March 2014 we took possession of most of a wing on the 3rd floor.

Starting from Scratch

Anyone who has set up a laboratory from scratch will appreciate the process we went through over the next six months as we worked to establish a functioning R&D capability. We needed to recruit staff, set up purchasing systems and then purchase everything we needed from pens, desks, chairs and lab-coats to spatulas, freezers, PCR machines, autoclaves, incubators and then arrange for their delivery and commissioning. Purchase of chemicals required establishing a register of all of the Material Safety Data Sheets and ensuring everyone was aware of the risks associated with each chemical use. This all necessitated establishment of laboratory procedures and standard operating procedures for use of equipment and materials. This is where the skills of a good laboratory manager become paramount. Even at this early stage it became obvious to me that from among our staff that Gary Devine had the necessary skills, attention to detail and commitment to this process and it earned him the right to both the title and the role.

Purchasing and installing everything from desks, computers, software and security systems, had to be organised and implemented as well as setting up routine storage, and conservation and management of laboratory cultures of Gd. All the while with new staff coming on board, there were HR matters to address, induction programmes and training before it was even possible to start discussing and organising everyone's personal research responsibilities and projects. And all to be done in the shortest possible time in order to ensure the laboratory could be up and running and delivering on our research priorities.

A pragmatic individual understands there has to be flexibility and allowances made to enable everything to settle into place. It is crazy to expect to get everything right, from the purchases made to the staff appointments. Things will always go wrong or not live up to expectations. Equipment might not meet specification or fit in the space available and in some cases it is necessary to lose individuals who do not fit the culture of the organisation. This latter point becomes quickly apparent in the first year when the pressure for everyone to play a part is crucial. Those individuals who are not adaptable or not suited to the frenetic world of a start-up soon become apparent. However, after 6 months things started to settle down and we were able to spend more and more of each day, and of each week concentrating on key scientific and R&D matters rather than just making everything function.

And there was so much to do!

Tricky TRIZ and Patent Protection

Technology companies are dependent on patent protection. Provided a patent that has been filed is eventually granted, it will provide up to 20 years of protection during which time the invention can be exploited by those who filed it, without competition in the territories covered by the patent. It is an expensive business paying for the maintenance of patent coverage in a number of countries so the technology has to be worth the expenditure. In other words, as soon as one takes out patents it is important to get into the market place and selling product while the protection from competition provides a market advantage.

For anyone who looks at patent publications, one of the obvious drawbacks that Azotic had was that there were few years left on the original patent before it expired in 2022, also the territories covered were not global. Despite Alan and Ted's best efforts the territorial patent coverage diminished each year to save the University patent costs. Hence, one of my main concerns, and that of the other Directors and of the investors, was that additional patents should be filed as soon as possible in order to extend the protection beyond 2022 when the University patent expired.

I have had the opportunity and unforgettable experience of patenting a couple of personal inventions during my life, namely a herbicide adjuvant and the means of rounding-up transactions at point of sales PIN terminals, as well as assisting on preparing a number of others. Defining an invention and determining whether it is patentable or not is an interesting and instructive process! It is all about ensuring you have invented something that has three attributes, firstly it has to be 'novel', then it has to be something that is not 'obvious to one skilled in the art' and lastly it has to have a genuine and measurable 'benefit or impact' of industrial relevance. In order to achieve all of these things one has to understand what has been discovered to date that is novel, what is already known or obvious to any skilled scientist and what information exists in the public domain (known as 'prior art'). As a scientist one tends to have a good feel for the scientific literature that relates to the subject matter of your interest, but knowing what has been done is a long way from knowing whether that knowledge impinges on the novelty or obviousness of your own invention.

In my position as CTO I was faced with the need to quickly identify opportunities that would generate new patentable inventions, it was useful to utilise a number of tools to identify where gaps existed in the patent landscape that would be advantageous for us to prioritise for study. One such technique is known as TRIZ analysis. TRIZ is the Russian acronym meaning the "theory of the resolution of invention-related tasks". I had come across the technique previously but had never had the chance to use it, but Karel Bolkmans was very familiar with it and encouraged us to explore what it could achieve for us. TRIZ essentially is a means of studying the patterns of invention in the global patent literature. Use of the technique has been extensively researched and it has generated some interesting insights into technical innovation. The three primary findings are that problems and their solutions, and how they evolve are repeated across industry sectors and different scientific disciplines, using innovation outside the field in which they were developed. This knowledge enables a user of TRIZ to create and to improve products, services and systems through identifying areas where gaps in innovation are most likely.

The TRIZ study that we undertook similarly provided us with a rationale for prioritising certain areas of research over others in terms of the likelihood that they would generate patentable inventions, filling gaps in the patent landscape. The TRIZ outputs provided the basis for designing the research programme to develop new patentable inventions based around Gd and its use.

Design and management of a research programme

The TRIZ analysis had provided us with the key areas of enquiry we needed to research that would most likely give us intellectual property that could potentially be patented. These key areas derived from the TRIZ analysis were (i) compatibility and or synergy with other microorganisms and agrochemicals, (ii) formulation and delivery methods and (iii) production systems, and as such provided the priority for our research programme.

Designing and carrying out research in a university is not the same as that required for a commercial company, especially a technology based start-up SME. The following provide a short list of some of those differences:

- Definitive time limited outputs that have demonstrable impact and add value to the company i.e. in this case delivers patents
- If a process or outcome is consistent, then it may not be necessary to know the underlying mechanisms
- There is not the luxury of being able to 'follow one's nose, or interests'
- Time is money, publications are useful if we have the time to write them
- Need for real team work, integration of effort and outputs

And yet scientists are free-thinking, independent, knowledgeable, opinionated and intelligent individuals who have been trained to solve the problems in their own way and to be sceptical of anything anyone else produces, or is not of interest to them and some at times might even be considered selfish, arrogant, argumentative and headstrong.

I have had the genuine privilege of managing teams of scientists for much of my scientific career, and I can safely say in terms of management, that they are one of the most difficult, opinionated, obtuse and complex groups of people to manage and lead. The term 'herding cats' is often used as a term to describe how difficult the process of managing scientists can be. But if you can make it work then they will deliver something that in artistic terms might almost be described as 'sublime', and that the difficulties of the process pale into insignificance relative to the outputs. Having forged a career since my 30s on research management I have tried, tested and published on relevant research management techniques that facilitate focus on common goals, relevant activities that deliver to integrated, cross-disciplinary and international outputs. With this experience, I was in a good position to know what needed to be done, but always it is really a case of creating an environment where everyone can not only achieve their personal goals, but they can do so in a way that is compatible with company goals. Often this becomes a case of a 'balancing jelly' and 'juggling custard' but few who have genuinely tried to get this right would ever boast about the extent of their achievement - because to have achieved it well, one's role should appear wholly inconsequential. Such is life!

I have always worked on the premise that scientists will undertake mundane tasks for a good proportion of their working day provided they always have some time to undertake research that really motivates and excites them. This is the balance that needs to be achieved within a commercial company that ensures the completion of routine sampling, monitoring, screening and evaluation with the delivery of more fundamental or innovative research, likely to generate new insights, opportunities, products and patents. Every industrial scientist needs a portfolio of projects and research experiments that appropriately spans this continuum. Without it, scientists become bored, uninspired, uncooperative and de-motivated. Hence, I ensured each scientist had a work programme that included both elemental and fundamental research and they could switch between different areas to ensure they could always be productive. This becomes important where a brick wall is hit, and no further progress can be made without further reflection, but research can still continue on different elements of a scientists' research programme, and keep things moving.

Something that is typical of an SME is the need to be able to remain highly flexible in delivering the day-to-day operations of a research programme. Fire-fighting to deliver an answer to a particular question that has come up from an investor or commercial partner has to be addressed as a priority and all hands may be required. This means resources, time and staff have to re-allocated, which in turn means someone's experiment may get put on hold or modified at short notice. Both from a scientific and a management point of view this can create major disruption and decrease the morale and effectiveness of the team - but it is just part and parcel of the working in such a dynamic and responsive research environment and a research manager has to put a lot of effort and energy into maintaining motivation and direction of the team.

Having said this, it was still a major exercise when the Board decided in 2016 that the priority was no longer on carrying out research for the purposes of patent applications but rather to use the laboratories to concentrate more on product development. It was not as if we were not already carrying out product development work, everything in an SME lab has to be delivering outputs relevant to product development, but it was a shift of emphasis that required winding down of some activities and prioritising a few new ones.

So after two years of working successfully in one sphere (4 patents were published from work in the period) it was now necessary to re-orientate output to follow the usual range of factors that have to be taken into account in developing a microbial product. Having gone through this process previously in developing a microbial insecticide based on the bacteria *Bacillus thuringiensis* and Green Muscle a myco-insecticide for locust control based on the fungus *Metarhizium anisopliae*, the territory was very familiar. So the team adapted, we re-issued projects and plans and the research took on a different look.

Product Development

The product development research took on a greater emphasis in the following areas:

- Bacteria stability and quality control
- Product formulation
- Product delivery
- Production and scale-up
- Shelf-life (formulated and on-seed)
- Diagnostics
- Compatibility with seed treatments and tank mixes of fertilisers

- Ecotoxicololgy - bacterial soil survival and seed transmission
- Packaging and labelling
- Regulatory matters

This is a varied and complex set of issues to address with such a small team of scientists, but month by month many, if not all the matters that needed to be addressed were addressed. Of course, they were sadly not all addressed with the positive outcomes one would have always liked, but for a large part great strides were made.

Publish or be damned

The academic community is sometimes faced with the issue of whether to publish a piece of research that has commercial potential or to keep the intellectual property confidential and apply for appropriate patent protection to enable commercialisation. The choice between publication and patents is an increasingly difficult dilemma for academics because on one hand their standard measure for career progression is the number of peer reviewed publications in high impact journals. If they delay journal publication in order to secure patent protection, it can potentially have a detrimental impact on their career. With the increasing requirement for academics to demonstrate 'impact' of their research however, the ability to demonstrate patentability of invention is also gaining higher value.

The situation is different in commercial research where developing a technology or product with associated patent protection is the primary driver. If this happens to have additional value in its publication as a peer reviewed scientific paper then preparation and submission of such work may be prioritised - or it may not.

Within Azotic Technologies we faced an additional complication because the NFix technology had the potential to be disruptive and challenged the prevailing paradigm. This meant that having the ability to publish in peer-review papers could be positive since it would potentially enhance our scientific credibility. However, the reality for commercial science is in prioritising competing and conflicting goals. To divert resources and effort to carry out any additional research that would be required to satisfy an academic publication was untenable from a commercial perspective, since the emphasis had to be on product development. In addition, since we needed to demonstrate field efficacy, the minimum requirement for an academic publication would be three seasons (years) of data, in the same crop in the same locations. Yet we were in a product development phase so we were not keeping variables constant across trials - we varied doses, concentrations of ingredients, manufacturing process, delivery method - all the time trying to refine our product. Until we settled on a single product, and then generated 3 years of data we would not have a coherent enough story for an academic publication.

I therefore decided that the best compromise would be to open our results as quickly as possible to technical scrutiny at scientific conferences and to publish a review of the scientific literature in order to collate what was known and present available evidence with regard to two things - demonstration of intracellularity of bacteria in plants other than Rhizobia and the ability of Gd to fix nitrogen *in planta*. They represent the two biggest criticisms of our approach.

It never ceases to amaze me how scientists who adhere to one particular paradigm, ignore or discount scientific research that does not substantiate their own limited view of the world. You will sometimes find in the literature research papers which provide evidence counter to the prevailing paradigm are ignored and not cited in publications. The other means of discounting their value or contribution is to undermine the perceived value of the research by adherence to a self-fulfilling system of peer review in high impact journals.

Hiding behind the anonymity of peer review is it too easy for scientists to prevent the publication of papers that are counter to their views, raising so many objections that it is impossible for them to be published. The more radical the discovery, and the greater its potential benefits, the more damaging the conflict between two paradigms may become.

I founded and was for 5 years chief editor of a review journal "Integrated Pest Management Reviews" and during that time experienced a number of instances where I had to over-rule the views of the referees. One in particular, a well-known and respected Professor who declared that it was 'dangerous' to publish a particular paper because the views held threatened mainstream thinking. My response had been that it would be a dangerous precedent 'not to publish' the paper exactly because it challenged the orthodoxy. The Professor concerned never again spoke to me or would have anything to do with me! I am not alone as an editor in such experiences.

Sadly this idea of totally objective, independent purely evidence-based scientists is actually counter to so much of the reality of the scientific world that I have encountered. Scientists are all too human, and the need to protect one's specialism in order to preserve their own research funding undermines and is counter to the greater needs of a nation for innovation. However, as Churchill is quoted as saying of democracy "Democracy is the worst form of Government except for all those other forms that have been tried from time to time". The current peer review system is flawed but it is still probably the best system open to us.

A Brief Review of Intracellularity

As mentioned in Chapter 3, back in 2006 Ted, Phil Stone and their colleague Mike Davey published their work on intracellularity of Gd in the journal of *In Vitro Cellular Development Biology - Plant*. However, this was not their first journal of choice - they had previously submitted the paper to *Nature* - the premier high impact scientific journal. The reviewers of the original submission to *Nature* (Ted was kind enough to show me the comments) were clearly greatly influenced by a background in rhizobial symbiosis. This is not surprising of course because the field of intracellularity of plant and nitrogen fixing bacterial symbiosis was dominated by the interaction between rhizobial and plant root nodules at that time. The possibility of intracellular symbiosis of another species of bacteria without the need for root nodules was hotly contested. This is in itself is a little bit surprising given that even back then there were cases of intracellular colonisation that did not involve root nodules or indeed Rhizobia. The most well-known and often cited is that of the intracellular colonisation of the cyanobacterium Nostoc in the leaf axis gland cells of the non-legume Gunnera. What evolution and biology is able to create once, it is possible for it to create again, not necessarily in the same way - as the rhizobial-nodule symbiosis and the nostoc/Gunnera leaf gland cell symbiosis demonstrates - and both systems facilitate nitrogen-fixation by their respective bacteria.

A comprehensive review of the scientific literature makes it apparent that there are other examples of intracellular colonising bacteria in plants some of which are also nitrogen fixers. Intracellularity has now been identified in peach palm (De Almeida *et al.*, 2009), Scots pine (Pirtilla *et al.*, 2000), Engleman spruce (Carrel & Frank, 2014), Limber pine (Carrel & Frank, 2014), banana (Thomas & Reddy, 2013; Thomas & Sekhar, 2014) and cactus (White *et al.*, 2012). Four of which include nitrogen-fixing bacteria (Carrel & Frank, 2014; White *et al.*, 2012). What has become clear is that root nodules are not an essential requirement either for the establishment of intracellular plant endosymbiosis or of nitrogen fixation.

The host plants cited above (peach palm, banana, coniferous trees and cacti) are all plant species that are capable of surviving in nutrient poor soils, and in evolutionary terms this would appear to make sense and provide an environment where plant fitness could be improved by selective pressure to form symbiotic relationships with nitrogen-fixing bacteria. Perhaps it is the case that such nitrogen-fixing symbiosis has been present all along but we have just not looked for it in the right places. We could take our hypothesis further, and Ted, Nick Gosman (R&D Manager from January 2017) and I have often mused over this, could it be that many crop plants also had intracellular associations with nitrogen-fixing bacteria but during their cultivation and domestication, and latterly intensive plant breeding under high nitrogen fertiliser regimes, we have just bred-out that association? We know for instance, that in Mauritius the natural symbiotic association of Gd with sugarcane has been bred-out of sugarcane varieties selected under high N conditions.

Nitrogen Fixation

Leaving aside the need for colonisation, intracellularity and the development of symbiosomes (considered above), BNF studies to demonstrate nitrogen fixation should look for increased benefits relative to controls associated with a number of measures. The ones that have been used with demonstrating or supporting hypotheses of nitrogen fixation in rhizobia include measures of foliage greenness/chlorophyll content, the percentage nitrogen found in the leaves, uptake of isotope $^{15}N_2$ through nitrogen fixation, presence and absence of nitrogen fixation with Gd mutants whose nitrogen fixing genes (Nif) are switched off, nitrogenase activity with acetylene reduction assays (ARA) and finally through the demonstration of field efficacy and yield benefits in nitrogen poor soils in the absence and/or presence of a nitrogen fertiliser.

From the existing scientific literature there are examples of Gd demonstrating in non-legumes all the same types of impacts achieved by Rhizobia in legumes. Chlorophyll levels have been demonstrated to increase in sorghum colonised by Gd (Bambara & Ndakidemi, 2009) and in *Lolium perenne* (Dent & Cocking, 2017). A 40% increase in leaf nitrogen has also been demonstrated in sorghum (Bambara & Ndakidemi, 2009).

For Gd ARA have successfully demonstrated nitrogen fixation of Gd in culture (Meenakshisundara and Santhaguru 2011; Madhaiyan *et al.*, 2004) and *in planta* (Momose *et al.*, 2013; Cocking *et al.*, 2006). The nitrogen isotope $^{15}N_2$ fixation activity has been used to successfully demonstrate nitrogen fixation with Gd (Momose *et al.*, 2013; Sevilla *et al.*, 2001; Madhaiyan *et al.*, 2004).

The ability to switch off the nitrogen fixing genes in a bacterium and thereby demonstrate the loss of capacity for the mutant to fix nitrogen is considered a key indicator of nitrogen fixation *in planta*. Research in this field has been published by Sevilla *et al.* (2001) demonstrating that shoot nitrogen content of sugarcane plants inoculated with wildtype Gd was significantly higher than control plants or plants inoculated with the *Nif*-mutants (lacking the ability to fix nitrogen) in nitrogen limiting conditions.

Crop yield benefits of Gd have been achieved in tomato plants (Luna *et al.*, 2012), Sugar beet (Abudureheman, 2012), sorghum and wheat (Luna *et al.*, 2010; Adriano-Anayal *et al.*, 2006).

Azotic demonstrated nitrogen fixation in its in-house R&D or by working with key research partners using all of the above techniques. However, we were always looking for new ways to demonstrate that Gd fixed nitrogen *in-planta* and it was Ted who suggested using a new technique, not previously used with a nitrogen fixing endophyte. The technique is known as NanoSIMs microscopy and at the time there were only two such microscopes in the UK, one at Oxford and the other at Manchester University. There are less than 25 such microscopes in the world. NanoSIMs stands for Nanoscale secondary ion mass spectrometry and is a nanoscale resolution chemical imaging mass spectrometer based on secondary ion mass spectrometry. For most of us this means little, however, what it does is identify the presence of an element, in our case nitrogen and through a colour image defines where and in what quantities the element is found within the plant. In 2017 Carvalho *et al.* (2017) first publicly reported the use of the NanoSIMs microscope with Gd colonised plants and was able to identify the presence of labelled ^{15}N levels in maize leaf chloroplasts due to biological nitrogen fixation by Gd. This is a major development and while this study was in its infancy, to be able to demonstrate nitrogen fixation has occurred in Gd colonised plants is an important breakthrough. Of particular importance is that the fixed nitrogen showed up in high concentrations in the chloroplast. Those remembering their basic biology of photosynthesis will recall that nitrogen is required to produce the plant pigment chlorophyll (which makes plant green and allows them to absorb light effectively) which is concentrated in the chloroplasts with leaf cells. Finding fixed nitrogen in such sites could go some way to explaining how, Gd colonised *Arabidopsis thaliana* plants showed higher whole canopy photosynthesis (De Souza *et al.*, 2015). The same study went on to show lower whole plant transpiration, and increased water-use efficiency. Essentially the colonisation by Gd of a plant could be increasing its overall efficiency of photosynthesis and this is one of the reasons we see an increase in yield. The attainable yield of such crops is being taken to a new level due to the presence of the bacteria.

Weight of Evidence

Part of the scientific process involves continuously reviewing the scientific literature to understand new developments as they arise. Another part of the scientific process, and one that perhaps occurs less often than it should, is to actively set about collating as a review the existing knowledge that has been published on different aspects of the problems that you are investigating. In my view there are too few reviews published and this is largely because there are no real career benefits in doing so. Academic careers are progressed through the publication of primary research papers rather than secondary publications such as reviews or books. In my opinion this completely misses the point of the perspective and insights gained by taking an overview. Having written a number of text books early in my scientific career it provided me with perspectives, insights and understanding of research across disciplines, of differing standards of proof and the means of integrating and of identifying opportunities for innovation at the margins of disciplines that a conventional reductionist silo-based approach could ever have delivered.

Hence, my default position is always to prepare a scientific review as a means by which I can challenge my own position or perceived wisdom and views on issues. It sometimes produces unexpected insights or perspectives that occasionally even materialise as a surprise. Within the BioCity research team I felt we had started to labour under a number of 'assumptions' that I thought it might be worth challenging and I naturally turned to the literature in order to expedite this process. Coincidentally at the same time, I was approached by the Open Publishing House - InTechOpen to write a chapter for a book that was being prepared on the subject of *Symbiosis*. Never one to miss an opportunity to create a genuine and useful output from a piece of research, I agreed to write the chapter. It also provided me with a deadline for my review, which if it had remained an internal document I may never have quite finished!

My chapter reviewing the scientific literature (Dent, 2018) added weight to our growing body of evidence that Gd is a truly 'extraordinary endophyte' nitrogen-fixing bacterium with important ancestral attributes.

Intracellular colonisation has for a long time been a source of contention concerning Ted's research on Gd and it is interesting that the intracellular environment, according to other researchers, is the main factor that correlates to genome size in bacteria (Toft & Andersen, 2010; Merhej *et al.*, 2009). In a study of the genomes of 350 species of bacteria comparing how closely their life history was associated with their host showed that genome size decreases the closer the bacteria life-history is intertwined with their host (Toft & Andersen, 2010). Those intracellular bacteria that cannot survive any part of their life-cycle without their host have genome sizes as small as 0.7 Mb, whereas those that are not wholly dependent on intracellularity have a median genome size of around 3.1 Mb with a whole range of life-history strategies in between. For bacteria that live and form associations with plants, then the range of genome sizes varies between 7.6 to 3.9 Mb with Gd having the smallest genome of the endophytes that have been studied to date. Interestingly this places Gd in the facultative intracellular coloniser category, having the ability to live both inter- and intracellularly within the plant but having very limited survival capacity outside the host, something Ted first demonstrated back in 2006 (Cocking *et al.*, 2006; Cocking, 2017).

Gd is also one of a few bacteria that has mechanisms that allow it to survive in the presence of very high levels of sugar (sucrose). High sugar concentrations can be used to preserve fruit in the form of jams and jellies simply because very few micro-organisms can survive in such an extreme environment. The ability of Gd to survive in high sugar environments is perhaps not too surprising given that its natural host plant is sugarcane, but one of the implications of high sugar concentrations is that it means there is limited water availability. Every bacterium requires water to survive and reproduce but Gd has an ability to survive in relatively low water to an extent greater than most other bacteria.

It this wasn't enough to make this bacterium 'extraordinary', it can also survive in a wide range of oxygen environments - able to respire and survive even when oxygen levels are really low. In re-reading about this for preparing the Chapter, I was reminded about the time I was introduced to this idea in 2015 and the person who raised it with us. As part of our efforts to create debate around non-nodular nitrogen fixation, Ted, Nathalie Narraidoo (an Azotic scientist) and I attended the 23rd North

American Conference on Symbiotic Nitrogen Fixation held at Ixtapa in Mexico back in December 2015. I had given a presentation entitled 'Establishment of Cereal and Non-Legume Symbiotic Nitrogen Fixation by *Glucoacetobacter diazotrophicus*' (Dent & Cocking, 2015) in which I had outlined the evidence for intracellular colonisation and explained the evolution of the NFix technology and the field results we were achieving at that time. To say that the response was muted from an audience that was made up of mostly scientists involved in rhizobial nitrogen fixation would be an understatement. However, at the end of the session I was approached by a man who introduced himself as Mauro Degli Eposti who calmly said to me that Gd was truly an extraordinary bacteria, and that it was perhaps even more extraordinary than I actually realised. With this enigmatic statement, the Professor then suggested I may like to hear more and suggested we meet with Ted in the hotel bar later and he would explain. I readily agreed.

Now aside from the fact that you may be beginning to feel that major turning points in this story seem to revolve around hotel bars, and wondering what that says about me, however, if we put that to one side - what we were actually able to learn from a conversation with Professor Eposti was truly illuminating and worth relating here. Professor Eposti was at that time on sabbatical at UNAM in Mexico, which by strange coincidence was the University from which Professor Caballero-Mellado had worked and collected the original strain of Gd that he sent to Ted. Professor Eposti explained that his research interests were with mitochondria (the organelle in each cell that manages respiration) and their ancestral origins.

For background, mitochondria are organelles that exist in all living cells (both plants and animals) that allow each cell to respire. Cellular respiration is the process of converting sugar, through the use of oxygen and its conversion into carbon dioxide to produce energy that can drive all of the functions of the cell. Ultimately it is what makes whole organisms, including humans, function. Mitochondria have their own DNA which is separate from genomic DNA and scientists generally believe this is because mitochondria were, in evolutionary time, originally an ancestral bacteria that was taken-up into cells. The search for these bacteria that were the early or 'proto'-mitochondria has been ongoing for many decades and Professor Eposti's research had led him to study the pathways and bio-energetic systems by which energy is produced. What the Professor went on to explain was the one of the outputs from his studies had shown that those bacteria that represented the closest match for the distal ancestors of mitochondria were a type of bacteria that are known as methylotrophs - but also his results indicated that the other most closely related bacteria that would fit the bill was a genus of bacteria that had some extraordinary attributes - namely *Gluconacetobacter* (Eposti *et al.*, 2014 a & b)!

If the genus of *Gluconacetobacter* has the ancestral attributes of mitochondria it raises interesting possibilities about the observed capability of Gd to move intracellularly and its ability to fix nitrogen in an oxygen 'rich' environment. Proto-mitochondria would also have required the ability to intracellularly colonise other cells in order to perform any respiratory function.

Research has shown that Gd has the ability to switch between situations of high and low oxygen concentrations and still function - which is not a common attribute among bacteria, and in order to do this Gd has to have special respiratory processes to cope. This is significant! Those who have argued against Gd fixing nitrogen intracellularly have based their argument on it having no means to control the destructive presence of oxygen on the nitrogenase enzymes. Rhizobia have a special haemoglobin-like protein called leghaemoglobin which absorbs the oxygen and allows the rhizobia to fix nitrogen. With no such similar mechanism in place it has been argued Gd cannot possibly fix nitrogen. However, the special respiratory chain and ancestral mitochondrial machinery of Gd enables it to respire at incredibly high rates thus potentially 'burning' up oxygen that may be present and restricting its impact on the sensitive nitrogenase enzymes. In doing so it also makes available high levels of energy required to facilitate nitrogen fixation. In addition, Gd has a large groups of genes associated with nitrogenase structure and function and a range of other mechanisms that protect the nitrogenase from oxygen. Hence it would seem feasible and it seems likely that, the bacterium combines these factors to enable symbiotic nitrogen fixation *in planta*. There is of course much more work that needs to be done here to elucidate such mechanisms but here we have the beginnings of an explanation for why Gd can do the things we understand it to be able to do in the laboratory and increasingly we're accruing evidence for in the field.

From my analysis and the research of others it is clear that Gd is a highly adaptive obligate endophyte, with a range of attributes, including intracellular colonisation capability; Gd certainly has the potential to fix nitrogen and thereby reduce nitrogen fertiliser use but whether this could be translated into positive and meaningful field trial results is the key to the future of the NFix technology.

The Good Guys Wear White

I grew up with a black and white television and the Saturday night family movie was, for so many of my formative years, a

Western. One thing that became clear is that in the Saturday Western, the bad guys always wore black and the good guys invariably wore white - or at least their hats were white! A similar phenomenon is evident for a later generation of movie-goers when the characters in Star Wars which also followed the white/black good/bad guy format. Were it so simple in real life!

In telling this story I wouldn't want you to think that everyone in this story is a metaphorical baddie in a black hat - because that certainly would not be true. But it has to be said the larger the potential for a technology to be disruptive, the more difficult it becomes for people not to be 'nay-sayers'. In fact it is easier and costs less in reputational risk to take a more sceptical stance. However, there are men and women in white hats out there, and bit-by-bit they emerge cautiously from the woodwork. It is important to point out that these are not individuals who are easily convinced, who readily have the 'wool pulled over their eyes' or who are not sceptical at all, but rather they are those people who have no vested interest, are prepared to stand apart from the crowd, to hold their scepticism in check, until their optimism and possibly hopes, are proven wrong rather than their scepticism proven correct. They tend therefore, to be prepared to share the risk and publicly offer support, in order to give the disruptive technology, its best chance of success. And if you think about it, why wouldn't everyone want that sort of outcome?

Along the way, scientists, government and intergovernmental representatives, environmentalists, climate change and sustainability advisers, philanthropists, businesses and farmers have chosen to share our journey and to offer advice and support, to look at the potential upside first and foremost and to take a risk. These are the people who collectively make it possible to deliver significant changes in the world.

A Few Awarding and Rewarding Moments

We gathered a number of those in white hats together (at that time a relatively small number) when we arranged an official opening of the Azotic BioCity laboratory. We felt that it was important to formally acknowledge the name of the laboratory for a number of reasons: (i) BioCity had proven to be an excellent place to establish our R&D capability, the staff and facilities support were first class, and under the leadership of Glen Cocker MBE and Toby Reid, the opening was a chance to publicly thank them for their support (ii) it was an opportunity to raise our profile within Nottingham and more widely, but most importantly (iii) it was an opportunity to recognise Ted's achievements and our respect for all that he had achieved.

Having known Sir John Beddington while he was the Government Chief Scientific Adviser through my work in Thailand, and having informed him and introduced him to Ted's research just the year before, I thought I would invite him to be our Guest of Honour for the event. John's availability had determined the date for the event which was set for the 5th June 2015.

Amid a fanfare of speeches and photographs we unveiled the name of our BioCity Laboratories, in honour of Ted as the 'Professor Ted Cocking Laboratories' in the presence of his family and our honoured guests. The plaque Ted was presented with stands at the entrance to the laboratories, proudly declaring and acknowledging the connection between Azotic and Ted Cocking, and it stands there still today.

As it happens, over the years we have not been the only ones who have been doing the honouring - we have been lucky enough and proud to be able to say that we were recognised for our achievements in developing NFix by a number of independent bodies. These included the:

- University of Nottingham Knowledge Exchange and Innovation Awards 2014
- Rushlight Awards 2014-2015: The Rushlight Award
- Frost and Sullivan 2015: European Alternative Fertiliser Technology Innovation Award
- New Energy and Cleantech Awards 2015: Innovator of the Year

But the one I was personally proudest to receive was from the Nottingham Post in 2017. A number of us went to the evening's event in Nottingham City to receive the Business Award for "Excellence in Science & Technology". These were good times as a team as we watched the awards grow.

7

TOO GOOD TO BE TRUE: THE "SNAKE OIL PHENOMENON"

The Honest Sceptic

Scepticism is a positive and necessary trait in science to ensure that research is properly and effectively scrutinised through the process of peer review. Healthy scepticism provides the sense of detachment that facilitates critical review. However, like any rather critical attribute, scepticism only has value in science provided it is not used cynically. A cynical use of scepticism can undermine objectivity and generate double standards of proof that then serve only to limit innovation, often only for the sake of a vested scientific interest. The use of disparaging tactics that undermine confidence in a novel technology may increase time to introduction into the market of a valid innovation; a delay that will ultimately serve no-one's long term interests and certainly can only undermine credibility in the scientific and innovation process, which is difficult and complex enough as it stands.

Disruptive technologies will often have the roughest of rides to market, as much because of vested interest in the market place on top of that from a scientific perspective. The very nature of a disruptive technology means that conventional approaches that have well-established markets will lose out to the new innovation. However, in the way that academics and researchers may be sceptical of paradigm shifting science, so industry is very wary of technologies that appear to be too good to be true. In industry the concern is always with the risk of being associated with the peddling of 'snake oil'.

Interestingly snake oil really does exist as a product, it is in fact a traditional Chinese medicine derived from a snake *Enhydris chinensis* which is applied as an ointment to relieve joint pain. The benefits of genuine snake oil were however, undermined particularly during the 18th and 19th Centuries, by dubious claims about products in the market. These fake products consisted largely of mineral oils or sometimes more hazardous materials and were sold by disreputable salesmen as a panacea for all ills. Hence, the commercial equivalent to the scepticism of paradigm shifting science for academics, are products that seem to be too good to be true, and are thus labelled as 'snake oil'!

While continued research by peers that elucidate mechanisms and the scrutiny of scientific publications will eventually win out for paradigm-shifting science, in the commercial world of agricultural there is a clear route of product validation in multinational, multiyear field trials. Even then however, it is, ultimately the performance and impact of a new technology in the farmers' fields that will eventually win over the sceptics.

The route to market for NFix then has required an extensive field trials programme. The pressure is always high for an SME to get products into the market not least because sales impacts on company value, but there are no short cuts in agriculture. Once a formulation and delivery method have been optimised it then usually takes a number of years of field trials under a range of different conditions, over a number of years in different countries for each crop. Thus in an ideal world, product formulations and delivery methods have been optimised for each crop, production system and planting machinery through highly controlled field trials prior to more commercial trials that relate more directly to real world farms and farmers. In reality, the commercial pressures are so great in an SME that the 'nice' distinction between the two phases is rarely applicable.

Trials with NFix were carried out on wheat and maize since these were the primary commercial markets as part of the product development pathway that led to commercial scale evaluation in 2018. In a way we were victims of our own early success, since we achieved such significant yield impacts in initial trials that time for refinement and optimisation became limited. Even so results for even what might be regarded as sub-optimal doses, formulations and delivery systems have proved incredibly positive generating highly significant yield increases that indicate major reductions in nitrogen fertiliser use are feasible without a yield loss.

Yield Responses and Nitrogen

Crop yield varies according to how much nitrogen is applied to the soil and it does this in a definable and repeatable way such that the more nitrogen that is added the greater the crop yield up to a point where the grain yield will plateau and even to decline if too much nitrogen is added. This type of response is typified by the curve shown in Figure 7.1, and is known as a yield-nitrogen response curve.

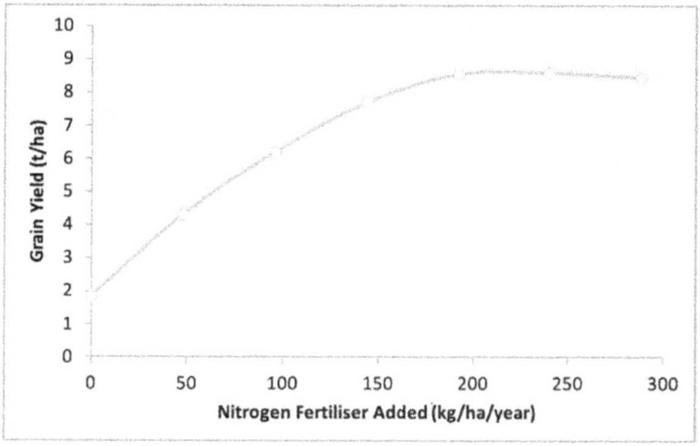

Figure 7.1a. Yield-nitrogen response curve for winter wheat (after Hawkesford, 2011).

Pedro Carvalho, our Chief Crop Scientist hypothesised that with the use of NFix the yield nitrogen response curve might be different from the standard curve with nitrogen fertilisers as depicted in Figure 7.1a. Pedro had two possibilities in mind. The first shown in Figure 7.1b shows how by comparison with the standard curve, the bacteria would fix nitrogen and have most impact on yield at the zero or low end levels of nitrogen fertiliser use. As the plant reached its 'normal' level of nitrogen required to produce the standard response the curve would level-off in line with the yields achieved with the optimum levels of nitrogen. This model assumes that the bacteria simply substitutes for mineral nitrogen fertiliser and provides no additional benefit to plant yield above that which can be provided by the mineral nitrogen at its optimum rate, e.g. at 200 kg/ha (Figure 7.1b).

However, Pedro also considered a further possibility, which involved the idea that the bacteria could actually provide an additional benefit, above and beyond that which the mineral nitrogen would provide on its own. If this were the case then the yield-nitrogen response curve might look something like that in Figure 7.1c. Where the attainable yield with NFix is higher than anything that can be achieved at any level of nitrogen fertiliser. This would imply the bacteria is able to provide additional benefits to plant growth and production efficiency. This would be ideal of course, but at this stage it was only a hypothesis.

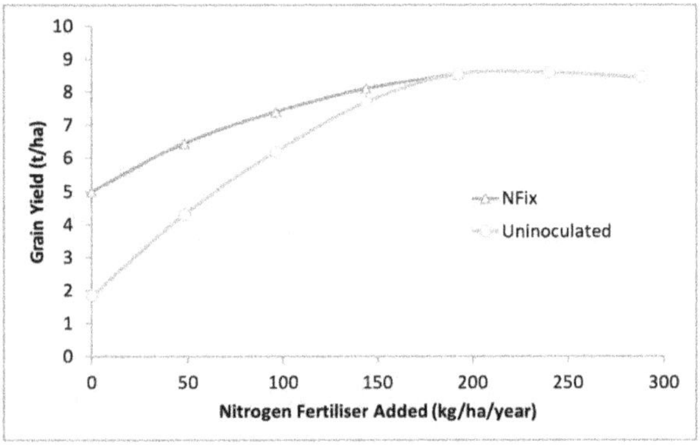

Figure 7.1b. The actual yield-nitrogen response after Hawkesford (2011) for winter wheat compared with a hypothetical yield response for NFix.

To achieve a yield-nitrogen response curve similar to Figure 7.1b would be fine, to achieve a response curve that in anyway mirrored the NFix response in Figure 7.1c would be truly remarkable and exceed all expectations. Such a response would suggest that the bacteria is not only fixing nitrogen but also having additional benefits, for example through promoting plant growth and possibly also photosynthesis.

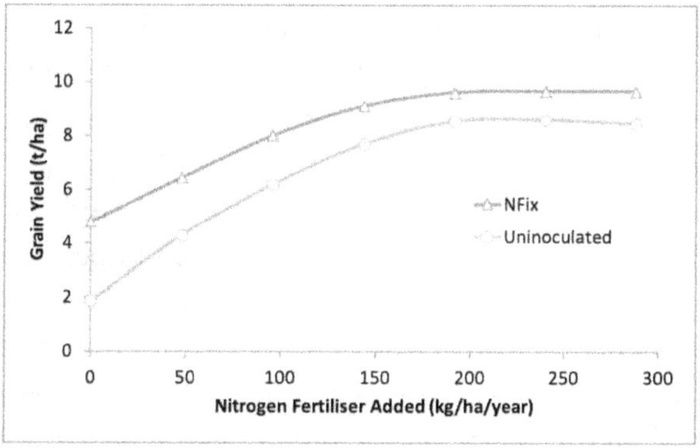

Figure 7.1c. The actual yield-nitrogen response after Hawkesford (2011) for winter wheat compared with a hypothetical yield response for NFix if the bacteria contributes more to plant growth than just nitrogen fixation.

If this is the case, as has been shown with the research plant species *Arabidopsis* (see Nitrogen Fixation, Chapter 6) it would suggest that the bacterial association with the plant is facilitating a whole new level of yield capability, above that which the plant can normally achieve on its own, even with the optimum level of nitrogen fertiliser applied.

Launching a fully-fledged programme of field trials for NFix was going to be a testing time. Under pressure as always to deliver an immediate benefit from my commercial colleagues waiting on the side-lines like brooding vultures searching for any signs that confirms their interest, we were unsure what the trials programme would deliver.

Generally, in plant breeding of crops such as wheat, a yield plateau has been reached where no more than a 2% yield increase seemed possible through conventional plant breeding techniques. If we could achieve something higher than this and reduce nitrogen fertiliser levels, then that might be considered a positive result. Alternatively, the agri-technology industry will only consider a new technology if it delivers something in the order of 5-6% increase in yield, so this might provide a better indicator of potential success, especially if we could achieve this and reduce nitrogen fertiliser levels. We were hopeful, but other than the initial trials in 2013 that had produced greater results than anticipated, expectations were probably running too high but we would see. Each season's crop trials would be carefully scrutinised. At this point we had still not had time to optimise our formulation for each crop and delivery system. These were testing times.

Maize

Maize (or corn in the USA) is a staple food crop in many parts of the world with total production higher than either wheat or rice. In 2017/18 the USA was the largest producer with crop production volume of around 371 million metric tons, with China and Brazil in second and third place for overall production (OECD/FAO, 2018). Interestingly, little of the maize produced is consumed directly by people, most of it is used for ethanol production, as animal feed or is processed to produce consumer products such as popcorn and corn syrup. Maize production is expected to increase in the next decade to meet a 16% increase in consumption by 2027 (OECD-FAO Agricultural Outlook 2018-2027;OECD/FAO, 2018).

Non-renewable energy consumption in USA maize production (from cradle-to-farm gate) ranges from 1.44 to 3.50 MJ/kg of maize, and GHG emissions associated with maize production range from –27 to 436 g CO_2 equivalent/kg of maize (Kim *et al.*, 2014). Most studies evaluated by Kim *et al.* (2014) show that soil N_2O emissions are the primary GHG source in maize production and account for 48–64 % of the total GHG emissions, 99–202 g CO_2 equivalent/kg, although N fertiliser production is also a primary GHG source in maize production (7–22 %). Hence, any reduction in the use of nitrogen fertiliser for maize could have a significant impact on nitrous oxide emissions from this global crop.

Azotic carried out a number of field trials in maize in Europe and the USA between 2014 and 2016. The crop seeds were treated with NFix prior to sowing, and then every other operational decision followed normal agronomic practice for the crop. What is important about the results presented here is that these are trials across three seasons and across two different continents. Most importantly is that the mean grain yields for the NFix treated crop plants is substantially higher than the controls with nitrogen fertiliser and critically follows a parallel curve across each level of nitrogen fertiliser use, as described by Pedro's hypothesis in Figure 7.1c! The bacteria is not only fixing nitrogen, it is also providing other benefits to the maize crop that is raising the overall attainable yield of the crop even increasing yield above the 100% recommended fertiliser rates (Figure 7.2).

The average yield impact across these ten trials is presented in Figure 7.2 and demonstrates an overall increase in yield of 8% (830 kg/ha). Firstly this yield increase is much larger than anything being achieved by plant breeding (around 2%) and higher than the minimum standard of 6% required by the industry for introduction of a new technology. From a scientific statistical stand point an estimation from a second order polynomial fit, predicts that NFix can replace 27% of the nitrogen fertiliser inputs without yield penalty (Carvalho *et al.*, 2017). So if a farmer wants to reduce the amount of fertiliser applied to his maize crop, he or she can do so by a quarter without suffering a decrease in yield.

If these sorts of results can be repeated or improved upon in other crops then the potential is enormous for NFix.

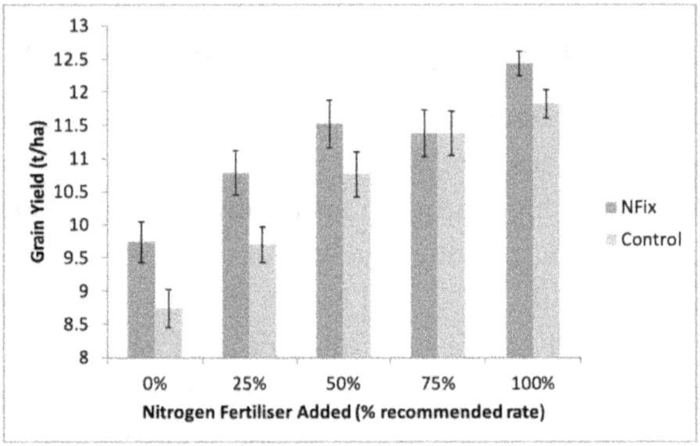

Figure 7.2. Mean grain yield data from 10 maize trials (2014: 4 Germany, 1 Belgium, 2015: 3 US, 2016: 2 US). The 100% fertiliser rate is determined from soil samples prior to sowing. Vertical bars are the standard errors of the means (after Carvalho *et al.*, 2017).

Wheat

The UN Food and Agriculture Organisation (FAO) in their agricultural statistics for 2018 quote the level of wheat production at 725 million tonnes. Wheat is grown on more land (ca. 215 million ha) and traded (US$50 billion) more than any other single crop. High in carbohydrates and protein (ca. 12% grain protein), wheat is consumed by 2.5 billion people in around 90 countries as bread, cakes, pastries, breakfast cereals, pasta, flour tortillas and noodles all made from wheat flour. Wheat is also used for cattle, poultry and other livestock feed.

Wheat is very sensitive to insufficient nitrogen and although N application has produced higher yields, this is not a linear relationship and there is an economic optimum application that needs to be determined for individual wheat cultivars (Foulkes *et al.*, 1998; King *et al.*, 2003). Nitrogen not only has a significant effect on crop yield but also on the protein content of the grain. High protein content is of nutritional benefit, particularly if coupled with the presence of a high content of essential amino acids (Shewry, 2007); additionally, high protein is required for optimum end-use quality, for example a minimum protein content of around 12% is required for bread making. Across most cultivars, there exists an inverse relationship between yield and grain protein (Simmonds, 1995), which means an inevitable consequence of increased yields appears to be decreased grain protein concentration, at least under constant N supply. In addition, nitrogen not only influences grain protein but also protein quality. All of this makes nitrogen a very important determinant of the value of a wheat crop.

A technology that not only allows for substitution of nitrogen fertiliser with a more sustainable alternative but also increases yields and protein content of grains would have a massive impact on climate smart wheat production. And this is because wheat is the largest user of global N fertiliser utilising 18.2%, compared to maize, the second largest user with 17.8% and then rice with 15.2% (IFA 2017). China (24.5%) leads the world in N fertiliser use in wheat, followed by India (16.5%), USA (11.5%) and Europe-28 (11.0%).

Although the results can vary according to soil type, production system and crop variety for example, the research of Gogoi and Baruah (2012) indicated that N_2O emissions from different wheat varieties ranged from 12 to 291 μg N_2O-N m^{-2} h^{-1} and seasonal N_2O emissions ranged from 312 to 385 mg N_2O-N m^{-2}. In a life cycle analysis carried out in Australia looking at the Carbon Footprint from one tonne of wheat Brock *et al.* (2012) demonstrated that 26% of emissions were derived directly from nitrous oxide from nitrogenous fertilisers applied to crop, but additional emissions occurred from CO_2 (up to 15%) and application machinery. This of course does not take into account the pre-farm production of nitrogen and transport which Brock *et al.* (2012) estimated to be responsible for 37% of the emissions (including those for lime). The important point here is that reducing nitrogen fertiliser use through use of a more sustainable alternative does not just make on-farm reductions in emissions but also in the production and transport.

For all of these reasons, wheat is an important crop for which a biofertiliser such as NFix could potentially deliver a significant positive impact. Field trials were conducted by Azotic to determine the extent of that benefit and although significant yield quantity enhancement was possible for spring wheat across three consecutive seasons (Figure 7.3) the situation was certainly more complicated and required further research to optimise systems for European winter sown crops.

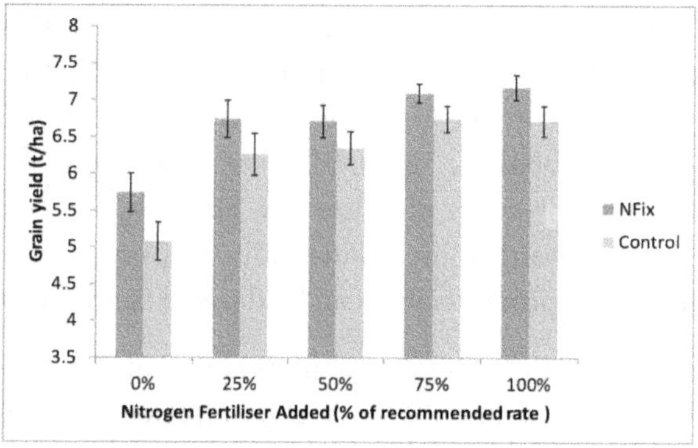

Figure 7.3. Yield against N fertiliser for spring wheat across locations in three consecutive years (2015 UK trials; 2016 UK trials; 2017 Germany and US trials) and N levels (Carvalho *et al.*, 2017).

The results in Figure 7.3 show how NFix inoculated seed increased yield by 7% (460 kg/ha) and demonstrated a staggering potential to reduce N-fertiliser applications by up to 61% with no reduction in yield. These results also demonstrate the synergistic effect of NFix and synthetic N fertiliser, leading to a yield increase above those achieved using the recommended fertiliser rate. Also savings in use of nitrogen fertiliser would provide an operational cost improvement for farmers as well as mitigating nitrous oxide green house gas emissions. This was a seriously important result and again we had demonstrated the yield impact across three years in different locations.

An additional and intriguing series of results demonstrated that even though yield quantity did not necessarily increase in trials in wheat, yield quality in terms percentage protein content did increase at all levels of nitrogen application (Figure 7.4). Protein results for all spring and winter sown and 2016 harvested wheat field trials generated a significant 4% increase in grain protein relative to uninoculated controls at all levels of N fertiliser. Such increase in protein levels are very impressive and significant from a milling and flour processing point of view. Indications are that improving bread-baking quality is certainly possible. High protein content of grain also has the advantage to farmers of increasing the grain value, potentially improving the grades achieved and the products that can be derived from the grain.

There is clearly more to learn and understand of the nature of the interaction of the bacteria with crop plants, not least wheat, and how this impacts on yield and grain quality. However, results presented here, demonstrate what is attainable and possible under field conditions. In this regard as field results they are very positive indeed.

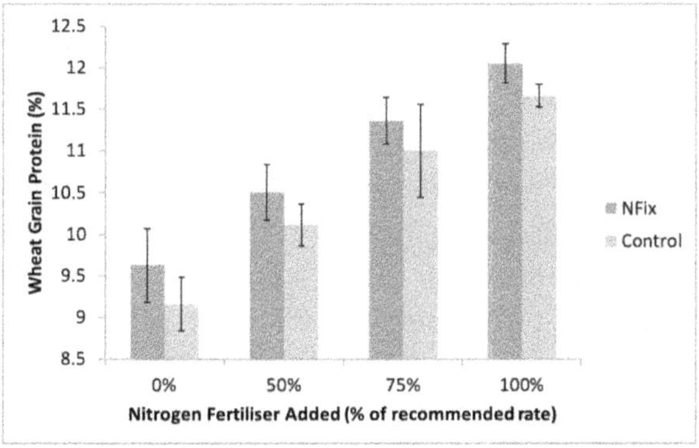

Figure 7.4. Combined yield data across all the 2016 UK (10), US (3) and Canada (2) wheat field trials (winter and spring sown), where NFix increased grain protein at all N fertiliser levels tested by 4% (after Carvalho *et al.*, 2017).

So in maize we had demonstrated we could increase yield-nitrogen response in the parallel curve in Pedro's second hypothesis (Figure 7.1c), the yield increase was on average 6% and if the farmer wished this could be traded for a 27% reduction in nitrogen fertiliser use. Then in Spring Wheat we had been able to demonstrate again the parallel yield-nitrogen response curve, an increase yield of 6% and a massive 65% reduction in nitrogen fertiliser use. In addition to this, the results showed we are able to increase grain protein content by 4% which on its own will have a significant impact on grain value for farmers and on flour and baking quality for the grain processors. Quite an achievement! But we were not finished yet.

Rice

Paddy rice production globally was in excess of 480 million metric tons in 2018 with China and India making the biggest contributions to production with 210 and 166 million metric tons respectively. Rice is the most widely consumed staple food crop worldwide and the most important grain with regard to nutrition and calorific intake. Application of nitrogen fertiliser has improved rice yields during the last 50 years but with considerable negative impacts on the environment (Huang *et al.*, 2017). From 1960 to 2012, the global N fertiliser consumption increased by 800%, and the annual N consumption in China increased from 8 to 35% of the world's N consumption (Wu *et al.*, 2016).

Currently, N_2O emissions from rice production is not widely or routinely monitored or reported and bizarrely often left out of greenhouse gas inventories from countries such as India and China (Kritee *et al.*, 2018). Policies on climate impacts of rice tend to assume that N_2O emission levels are negligible or small at <10% of the total CO_2e100^y, even under intermittently flooded conditions (CCAFS, 2017; Richards & Sander, 2014), whereas there is some evidence to suggest that, rice-N_2O contributes 25% to the GHG impact of rice cultivation on a CO_2e100^y basis (EPA, 2013; Myhre *et al.*, 2013).

Methane from global rice production currently accounts of a half of all crop-related GHG emissions. Several international organisations are advocating reduction in methane emissions from rice by promoting intermittent flooding without accounting for the possibility of large emissions of nitrous oxide (N_2O), a long-lived GHG. Nitrous oxide emissions from intermittently flooded rice fields could be 30-45 times higher than reported under continuous flooding. Net climate impacts of rice cultivation could be reduced by up to 90% through co-management of water, nitrogen and carbon. To do this effectively will require a careful ongoing global assessment of nitrous oxide emissions from rice - or we will be ignoring a very large source of climate impact (Kritee *et al.*, 2018).

The ability to deliver nitrogen to rice plants via bacterial nitrogen fixation would remove this dilemma, at least if sufficient nitrogen fertiliser can be replaced by use of NFix.

Ted had worked with rice early on in his search for intracellularity with Rhizobia and had demonstrated that it was possible to induce a 'nodule-like structure' in rice roots, but alas no significant nitrogen fixation was detected. Rice was also one of the crops Ted had initially demonstrated it was possible to achieve intracellular colonisation with Gd (Cocking *et al.*, 2006). Hence, there was good reason to believe that rice would be a good candidate species for use with NFix. This view was enhanced further by many paddy rice production systems that involve pre-germination of seed in water prior to field planting. Such a system would potentially allow for inoculation of the rice roots in a relatively controlled environment prior to planting in the field. Given the bacteria can survive well in water and that moisture is required by the Gd to facilitate movement, being able to inoculate an early germinating rice root should provide an ideal environment for Gd colonisation. Recognising this opportunity very early on in 2014, I suggested rice might be a good target crop to my colleagues. At the time I was still a member of the Thai Government International Advisory Board on Science, Technology and Innovation, so I had the opportunity to discuss and seek advice from scientific colleagues in Thailand. There was agreement that the technology had tremendous potential and relevance to rice systems in Thailand, and colleagues in BIOTEC put together a draft collaborative programme for research. On the same visit I also had the chance to meet with a few key rice seed distribution companies who were very interested in the possible commercial testing of a product.

Despite an obvious fit between technology, the potential delivery system and the rice cropping system, in-country research and potential commercial partners, the Azotic Board decided that rice should not be a priority at that time which was, to say the least, disappointing. However, I was given the go-ahead to evaluate NFix closer to home in Spanish rice product systems, which we pursued at a very low level and rather surprisingly with no real significant success. This was perplexing and difficult to rationalize and it was not until our Koppert colleagues tested NFix in Vietnam in 2016 that a real break-through in rice was achieved. An optimized dose, formulation and delivery system emerged from further laboratory and field evaluation that was subsequently validated in three countries - Vietnam, Thailand and the Philippines.

The formulation and delivery method was demonstrated to fit well with existing seed priming systems and demonstrated that NFix an reduce necessary nitrogen fertiliser levels while maintaining yields and reducing farmers operating costs in each of the three rice producing countries where it was tested.

As Azotic's published results on rice demonstrate (www.azotictechnologies.com):

1. Overall the NFix treated rice yields were significantly higher ($p \leq 0.001$) than the untreated at every N level tested.
2. Across all of the trials at all N levels the NFix treated plots yielded at least 17% more than the untreated, with a 1 t/ha (29%) increase being seen at 50% N.
3. The maximum yield increases observed in individual trials were 34% at 0% N, 57% at 50% N and 15% at 100% N.

These are very positive results indeed that clearly demonstrate the potential of NFix in rice, significantly improving attainable yield increases whist reducing N fertiliser applications by up to 50% (Figure 7.5). In addition, when NFix s combined with the full-recommended rate of N fertiliser then a significant yield increase of around 7% is achieved. This is equivalent to 0.3 t/ha and was observed in both the Thailand and the Philippines field trials. The importance of this is that it yet again demonstrates that Gd's nitrogen-fixing capability is not inhibited by the presence of the nitrogen fertiliser. The product delivers an additive yield impact even at full-recommended rates of nitrogen fertiliser. While this may not be optimal from the point of view of climate smart agriculture, it may certainly address some concerns over future food security.

In the three trials in Vietnam, the overall response for NFix as a mean average 19% yield increase while in Thailand and the Philippines the NFix treated plots yielded 17% more than the untreated. These yield results are phenomenal and when combined with the climate benefits that can be accrued through reduced use of nitrogen fertilisers, the future benefits and use of this technology are assured.

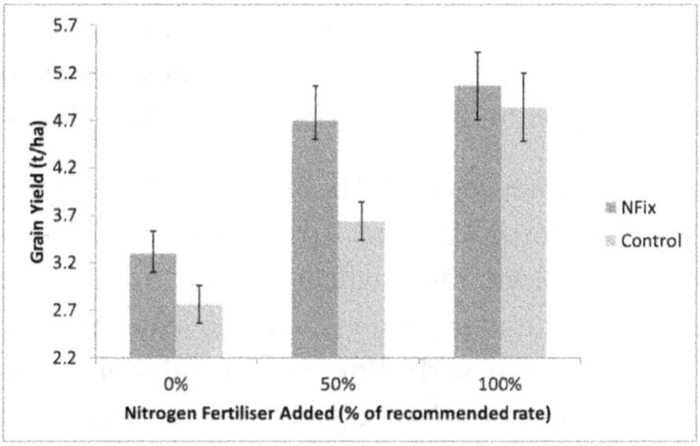

Figure 7.5. Average rice yields (t/ha) across the seven NFix treated and untreated trials during the 2017/18 growing season in Vietnam (3 trials), Thailand (2 trials) and the Philippines (2 trials) (derived from Azotic 2018, www.azotictechnologies.com).

Prospects for changing agriculture

As these field trial results continued to accumulate, I have had the pleasure of being able to report at numerous international conferences the progress that has been made in the development and evaluation of NFix.

One such key development has been the ever-expanding list of crops in which the Azotic scientists have been able to demonstrate Gd colonisation. The Azotic strain has been shown to colonise crop plants as diverse as grass (*Lolium perenne*) tea, coffee, potato, oil palm, tomato, rice, wheat and maize, barley and oil seed rape (Dent, 2016; Dent, 2017 b & c), as well as several others, which for a species that has been largely associated with sugarcane - is quite remarkable. Such findings also demonstrate the potential scope of use of NFix and Gd for improving crop production globally. If yield results such as those reported for wheat, maize and rice can be replicated in these other crops then the potential impact for global agriculture could be immense.

Since 2013, Azotic has tested NFix in 224 yield nitrogen response trials in Europe, North America, Africa and Asia on the following crops: grasses, oilseed rape, wheat, barley, maize, rice, potato, tomato, sugar beet and soybean (Carvalho *et al.*, 2017). The trials have been carried out with different formulations and delivery systems in order to prioritise crops and markets relevant to initial commercialisation.

A surprisingly field trial result was achieved with Soybean during 2017. Soybean is a legume so has its own root nodule associated Rhizobia to fix nitrogen. Ted Cocking had previously demonstrated that Gd would colonise legumes in his study with white clover (Cocking *et al.*, 2006) but it was only hopefully anticipated that Gd would impact at all on soybean growth in the presence of Rhizobia and root nodule nitrogen fixation in the field. It is early days but, as an indicator of potential, we were able to demonstrate under field trial conditions a highly significant yield increases of 29% at zero N and 8.4% at 44.8 kgN/ha when using NFix compared to the untreated control in a US field trial.

Azotic has also been carrying out research with potatoes and generating promising results. But whether it is soybean, potato, cereal or oil crops the important point on which to focus is that the widespread potential of NFix and related products. The ability of Gd to colonise, to fix nitrogen, to impact both on yield and nitrogen fertiliser reduction is enormous and the research to date has demonstrated this quite convincingly. But this is only the beginning of a journey!

8

WHERE THERE IS A BEGINNING ... THERE IS SOMETIMES - NO END.

In six years the Azotic R&D team and a few key partners have been able to take what was a laboratory proven concept, to develop an initial inoculant formulation and demonstrate field efficacy beyond any level of expectation in terms of yield benefits and nitrogen fertiliser reduction. It has been possible to do this in crops as diverse as wheat, rice, maize, and turn it into a field-validated technology relevant to key commercial target markets.

Commercially Available

On December 7th 2018 Azotic announced in the agricultural trade journal *Agropages* (http://news.agropages.com/News/News-Detail---28632.htm) its intentions to make its NFix technology, branded as Envita™ in the USA, commercially available. The article stated "After positive field trial results on maize, soybean and rice, Envita is now commercially available to growers in 26

US States. In 2018, farm scale trials across 11 states validated the previous 6 years of research with encouraging results and feedback from growers." Nolan Berg, President of Azotic North America, went on to say "Envita increases yields on average 5-13% and in some cases up to 20% in trials where nitrogen fertiliser levels have not been reduced."

This is a remarkable achievement within six years from a standing start in a University laboratory to a farm-scale evaluated product and commercialisation for such a disruptive and potentially revolutionary agricultural technology. However, it is not to say that there is not further research and development that still needs to be done. What the process of getting the technology this far, of gaining greater understanding of, and developing Gd into this initial product, has shown is that there is so much more potential yet to exploit within this organism using the intracellular endophytic approach. There are still many unknowns but also many 'knowns' that can be developed and exploited in order to improve production, formulation, storage and delivery of this remarkable bacterium.

A Growing Need

The conundrum we face is the need to provide food for a growing world population while mitigating against the use of agricultural practises that remain the biggest emitter of harmful greenhouse gases contributing to climate change. We have no choice in these matters we have to find solutions or we face widespread future famine, particularly in Africa, and with the projected increase in global temperature above the critical 2 degrees, a global climate change catastrophe. It is imperative to reduce nitrous oxide gas emissions from fertiliser production and use if we are to have any hope of mitigating against agricultural contributions to climate change.

Famine prevention and food security necessitate a staggering increase in staple food crop yields in commercial cereal production (wheat, rice, maize) but also those crops such as cassava, plantain, yams, sweet potato that sustain millions of small-holder farmers in developing countries. Both commercial and small-holder farmers face constraints to increasing crop yields in ways that will not contribute towards increased GHG emissions. For commercial farmers they need sustainable alternatives to synthetic nitrogen fertilisers and small-holder farmers, who have historically struggled to gain access to synthetic nitrogen fertilisers (cost, logistics and lack of infrastructure, often providing barriers to uptake), need an alternative that is both available and accessible. In both cases the response to this challenge are nitrogen-fixing bacteria such as Gd formulated in appropriate ways to facilitate delivery and application to the full range of commercial and subsistence crops across the world. This is not about genetic engineering and genetically modified organisms but about harnessing the existing ingenuity of naturally occurring populations of organisms like Gd that will deliver the 'Good Nitrogen' that crop plants require. It can be done!

The situation has so additional potential because there is good evidence to suggest that some crop cultivars are more susceptible to colonisation by Gd than others. The implications of this is that increased colonisation of nitrogen-fixing bacteria can be enhanced through crop breeding methods - using existing conventional plant breeding techniques (non GMO) and infrastructure. Seed might even be delivered with the bacteria already inside, having been taken-up in the developing seed from the parent plant. These are not pipe dreams; the available scientific evidence already indicates that this is possible but just not occurring regularly enough to be of value - but plant breeding could resolve that problem.

The world is too slowly adapting to the need to drastically reduce GHG emissions sufficiently to avoid the critical 2 degree rise in temperature. In no other sector is this more true than in agriculture. Other sectors such as transport and energy have been making significant advances towards more sustainable alternatives to current practices, but agriculture is lagging far behind when it comes to reducing GHG emissions. The situation is such that effectively business as usual in agriculture will see an increase of its proportional GHG contribution - which is somewhere around 25-30% at the moment, rise even further! Agriculture has become a major climate change issue - and nitrogen is a significant component of this.

Reducing the need for ammonia production and its use by farmers because crop yields can be maintained using NFix can have a massive impact on GHG emissions. If for example Haber-Bosch ammonia production can be reduced by 25% because of reduced farmer demand for fertilisers then this would, on its own reduce carbon emissions by around 280 million tonnes. That same reduced use of nitrogen fertiliser would potentially lower nitrous oxide gas emissions from the field by a further 300,000 tonnes as a conservative estimate - which converts to 89 million tonnes of carbon. Thus a potential total of 369.4 million tonnes of carbon would be removed from use of NFix replacing just 25% of nitrogen fertilisers globally. These are gross estimates but demonstrate the potential and if this can be improved upon even further then ...?

With the delivery of products based on the intracellular colonisation of Gd now available it is possible to imagine a world where a small nitrogen-fixing bacterium produced *en masse* and delivered as a seed treatment and/or with the help of an international plant breeding programme is able to avert the worst excesses of global climate change and famine. We have a genuine solution on offer to resolve our climate and food security conundrum. We have to ask ourselves, that if this is even only a potential possibility and the solution is not yet perfect, why wouldn't we want to allocate resources and effort to making something so extraordinary a reality? With a real life example of the bacterium - Gd, having demonstrated the ability to deliver this G*oo*d Nitrogen in farmers fields, what excuse do we have not to pursue things further?

The Sustainable Nitrogen Foundation

Alan Burbidge and I met in a bar in Bangkok on 8th March 2011 and we then met with Ted at the Lakeside Café on Nottingham University campus just 1 month later. From those meetings our journey started but it is a journey that has not yet ended.

Ted Cocking, Alan Burbidge and I have established the Sustainable Nitrogen Foundation (Charity Registration Number 1180511: http://www.tsnf.org.uk) with the objective of continuing our journey to ensure this NFix technology or subsequent versions of it, has a global impact – not least for small holder subsistence farming and developing country agriculture.

The object of the Foundation is to advance environmental protection or improvement for the benefit of the public by carrying out scientific research into biological nitrogen fixation and publishing the useful results, with a view to securing outcomes in the mitigation of greenhouse gas emissions and water pollution derived from nitrogen fertilisers in agriculture.

A new phase of the journey of taking intracellular nitrogen fixation with *Gluconacetobacter diazotrophicus* now begins and there is definitely a great future ahead for its development and use globally - with no end yet in sight. We need partners and supporters to make this possible and to realise the vision we have for a sustainable greener agriculture with nitrogen fixation for all crops. If this is a vision and a mission that inspires you, if this is something to which you can commit to ensure food security and reduced nitrogen pollution - then join with Ted, Alan and I to make something truly extraordinary happen!

REFERENCES

Abudureheman A (2012) Improving sugar beet productivity by inoculation with *Gluconacetobacter* spp. M.Sc. Thesis, Saint Mary's University, Halifax, Nova Scotia. 2012.

Adriano-Anayal M, Salvador-Figueroa M, Ocampo J A, and García-Romera I (2006) Hydrolytic enzyme activities in maize and sorghum roots inoculated with *Gluconacetobacter diazotrophicus* and *Glomus intraradices*. *Soil Biol. Biochem.* 38, 879–86.

de Almeida C V, Andreote F D, Yara R, Tanaka F A O, Azevedo J L, de Almeida M (2009) Bacteriosomes in axenic plants: endophytes as stable endosymbionts. *J Microbiol. Biotech.* 25, 1757-1764.

Attwood M M, Van Dijken J P and Pronk J T (1991) Metabolism and Gluconic Acid Production by *Acetobacter diazotrophicus*. *J. Ferment. Bioeng.* 72 (2), 101-105.

Bambara S and Ndakidemi P A (2009) Effects of inoculation, lime and molybdenum on photosynthesis and chlorophyll content of *Phaseolus vulgaris*. *Afr J Microbiol. Res.* 3 (11), 791–8.

De Bono E (1970) Lateral Thinking: a Textbook of Creativity. Penguin Books, Aylesbury, Buckinghamshire.

Brock P, Madden P, Schwenke G and Herridge D F (2012) Greenhouse gas emissions profile for 1 tonne of wheat produced in Central Zone (East) New South Wales: A life cycle assessment approach. *Crop Past. Sci.* June 2012. DOI: 10.1071/CP11191.

Carrel A A and Frank A C (2014) *Pinus flexilis* and *Picea engelmannii* share a simple and consistent needle endophyte microbiota with a potential role in nitrogen fixation. *Front. Microbiol.* 5 (333), 1-11.

Carvalho P, Narraidoo N, Gosman N, Cocking E C and Dent D (2017) *Gluconacetobacter diazotrophicus*: delivering a more sustainable Wheat and Maize yield. Poster at the ICNF Granada September 2017. Available from http//www.azotictechnologies.-com. [Accessed: 2017-12-21].

Cavalcante V A and Döbereiner J (1988) A new acid-tolerant nitrogen-fixing bacterium associated with sugarcane. *Plant Soil* 108, 23-31.

Cocking E C, Stone P J and Davey M R (2006) Intracellular colonisation of roots of Arabidopsis and crop plants by *Gluconacetobacter diazotrophicus. In Vitro Cell Dev. Biol. Plant.* 42(1), 74-82.

Cocking E C (2017) The Greener Nitrogen Revolution: Cereal and Other Non-Legume Crop Symbiotic Nitrogen Fixation. Proceedings 814 Paper presented to the International Fertiliser Society Conference in Cambridge, UK, 8th December 2017. ISBN 978-0-85310-nnn-n.

CCAFS (2017) Mitigation strategies in rice production, in collaboration with the Climate and Clean Air Coalition (CCAC). Low Emissions Development Research Flagship, The CGIAR

Research Program on Climate Change Agriculture Food Security (Wageningen University and Research, Wageningen, The Netherlands).

CCC (2017) The CCC Meeting Carbon Budgets: Closing the policy gap 2017 Report to Parliament. Committee on Climate Change June 2017. Committee on Climate Change Copyright 2017.

Dent D R (1990) *Bacillus thuringiensis* for the control of *Heliothis armigera*: Bridging the gap between the laboratory and the field. *Aspects Appl. Biol.* 24,179-185.

Dent D R (1991) Insect Pest Management. CABI Wallingford. 604.

Dent D R (1995) Experimental Paradigms. In: Dent D (ed) Integrated Pest Management. Chapman and Hall London. 172-208.

Dent D R (2013) Nitrogen fixation in corn and 199 other crops. Presentation at the Ontario Agri-Food conference in Guelph, Canada in April 7th 2013.

Dent D R (2015) Promoting Plant Growth through Nitrogen Fixation. Presented at seminar by special invitation to Campden BRI Industry Members 27th May 2015.

Dent DR (2016) Boosting Crop Production with Natural Nitrogen. Presentation at New Frontiers in Crop Research SCI London Thursday 20 October 2016.

Dent D R (2017a) Low Carbon Opportunities in the Agricultural Sector. Presentation at European Parliament, Brussels, Belgium, 25th January 2017.

Dent D R (2017b) Translating Laboratory and In-field Efficacy Data for Successful Product Design. Presentation at the Microbiome AgBiotech Summit: Harness the Power of Microbes for innovative, Productive and Sustainable Agricultural Practice. February 21-23 2017, Raleigh, North Carolina, USA.

Dent D R (2017c) Translating Laboratory and In-field Efficacy Data for Successful Product Design. Presentation at the Microbiome AgBiotech Europe: Harness the Power of Microbes to Deliver Effective and Consistent Agricultural Products of the Future. 20-21st September 2017, London, UK.

Dent D R (2018) Non-nodular Endophytic Bacterial Symbiosis and the Nitrogen Fixation of Gluconacetobacter diazotrophicus. Chapter 4 in Symbiosis. Ed Everlon Cid Rigobelo *Intec Open.* 53-81.

Dent D R and Cocking E C (2015) Establishment of cereal and non-legume crop symbiotic nitrogen fixation by *Gluconacetobacter diazotrophicus.* Presentation at the 23rd North American on Symbiotic Nitrogen Fixation Conference Ixtapa Mexico, Dec 6-10th 2015.

Dent D R and Cocking E C (2017) Establishing symbiotic nitrogen fixation in cereals and other non-legume crops: The Greener Nitrogen Revolution. *Agric. & Food Sec. 6, 7 DOI 10.1186/s40066-016-0084-2.*

Dobermann A. (2007) in Fertiliser Best Management Practices: General Principles, Strategy for their Adoption and Voluntary Initiatives vs Regulations. *International Fertiliser Industry Association, 2007.* 1–28

EPA (2013) Global Mitigation of Non-CO_2 Greenhouse Gases: 2010–2030 (United States Environmental Protection Agency, Office of Atmospheric Programs, Washington, DC), pp 19–42.

Esposti M, Chouaia B, Comandatore F, Crotti E, Sassera D, Lievens P M-J, Daffonchio D and Bandi C (2014) Evolution of

Mitochondria Reconstructed from the Energy Metabolism of Living Bacteria. *PLOS ONE.* 9(5):1-22. DOI: 10.1371/journal.pone.0096566.

Esposti M (2014) Bioenergetic Evolution in Proteobacteria and Mitochondria. *Genome Biol. Evol.* 6(12), 3238–3251. DOI:10.1093/gbe/evu257.

Erisman J W, Sutton M A, Galloway J, Klimont Z and Winiwarter W (2008) How a century of ammonia synthesis changed the world. *Nature Geosci.* 1, 636–639.

Flynn H C and Smith P. (2010) *Greenhouse Gas Budgets of Crop Production – Current and likely Future Trends* First edn (IFA, 2010).

Foulkes M J, Sylvester-Bradley R and Scott R K (1998) Evidence for differences between winter wheat cultivars in acquisition of soil mineral nitrogen and uptake and utilisation of applied fertiliser nitrogen. *J. Agric. Sci.* 130, 29–44.

Fryzuk M D (2004) Inorganic Chemistry: ammonia transformed. *Nature* 427 (6974), 498-9.

Van Grinsven H, Ward M H, Benjamin N and de Kok T M C M (2006). Does the evidence about health risks associated with nitrate ingestion warrant an increase of the nitrate standard for drinking water. *Env. Health.* 5, 5-26.

Del Grosso S J. and Grant D W (2011) Reducing agricultural GHG emissions: role of biotechnology, organic systems and consumer behaviour. *Carbon Manag.* 2, 505–508.

Global Rice Science Partnership (2013) Rice Almanac (International Rice Research Institute, Los Baños, Philippines), 4th Ed.

Huang S, Zhao C, Zhang Y and Wange C (2018) Nitrogen Use Efficiency in Rice. Chapter 10 in Nitrogen in Agriculture – Updates. Intech Open p 187- 208. http://dx.doi.org/10.5772/intechopen.69052. Springer ISSN 1939-1234.

IFA (2017) Assessment of Fertiliser Use by Crop at the Global Level 2014/15. A/17/134 rev. November 2017. International Fertiliser Association and International Plant Nutrition Institute. 1-18.

Kim S, Dale B E and Keck P (2014) Energy requirements and Greenhouse Gas Emissions of Maize Production in the USA. *BioEnergy Research* DOI 10.1007/s12155-013-9399-z.

King J, Gay A, Sylvester-Bradley R, Bingham I, Foulkes J, Gregory P and Robinson D (2003) Modelling cereal root systems for water and nitrogen capture: towards an economic optimum. *Ann. Bot.* 91, 383–390.

Kritee K, Nair D, Zavala-Araiza D, Proville J, Rudek J, Adhya T K, Loecke T, Esteves T, Balireddygari S, Dava O, Ram K, Abhilash S R, Madasamy M, Dokka R V, Anandaraj D, Athiyaman D, Reddy M, Ahuja R and Hamburg S P (2018) High nitrous oxide fluxes from rice indicate the need to manage water for both long- and short-term climate impacts . *PNAS* 115 (39), 9720–9725.

Kuhn T S (1962) The Structure of Scientific Revolutions. *International Encyclopedia of Unified Science*. 2 (2), 210.

Luna M F, Aprea J, Crespo J M and Boiard J S (2012) Colonisation and yield promotion of tomato by *Gluconacetobacter diazotrophicus*. *Appl. Soil Ecol.* 61, 225–9.

Luna M F, Galar M L, Aprea J, Molinari M L and Bioardi J L (2010) Colonisation of sorghum and wheat by seed inoculation with *Gluconacetobacter diazotrophicus*. *Biotechnol. Lett.* 32, 1071–6.

Madhaiyan M, Saravanan V S, Jovi D B, Lee H, Thenmozhi R, Hari K, *et al.* (2004) Occurrence of *Gluconacetobacter diazotrophicus* in tropical and subtropical plants of Western Ghats, India. *Microbiol. Res.* 159 (3), 233–43.

McCaffrey D (2011) Raw sugarcane juice – nature's perfect wonder food. Centre for Processed-Free Living. http//: www.processed-freeamerica.org.

Medawar P B (1981) Advice to a Young Scientist. Pan Books, London.

Meenakshisundara M and Santhaguru K (2011) Studies on association of arbuscular mycorrhizal fungi with *Gluconacetobacter diazotrophicus* and its effect on improvement of Sorghum bicolour. *Int J Curr Sci Res.* 1 (2), 23–30.

Merhej V, Royer-Carenzi M, Pontarotti P and Raoul D (2009) Massive comparative genomic analysis reveals convergent evolution of specialised bacteria. *Biol Direct* 4, 13 DOI:10.1186/1745-6150-4-13

Momose A, Hiyama T, Nishimura K, Ishizaki N, Ishikawa S, Yamamoto M, *et al.* (2013) Characteristics of nitrogen fixation and nitrogen release from diazotrophic endophytes isolated from sugarcane stems. *Bull. Fac. Agric. Niigata Univ.* 66 (1), 1–9.

Muthukumarasamy R., Revathi G. and Loganathan P (2002) Effect of inorganic N on the population, *in vitro* colonisation and morphology of *Acetobacter diazotrophicus* (syn. *Gluconacetobacter diazotrophicus*). Plant Soil 243, 91- 102.

Myhre G, Shindell D, Bréon F -M, Collins W, Fuglestvedt J, Huang J, Koch D, Lamarque J -F, Lee D, Mendoza B, Nakajima T, Robock A, Stephens G, Takemura T and Zhang H (2013) Anthropogenic and Natural Radiative Forcing. In: *Climate Change 2013: The Physical Science Basis. Contribution of Working Group I to the Fifth Assessment Report of the Intergovernmental Panel on Climate*

Change. Eds. Stocker T F, Qin D, Plattner G -K, Tignor M, Allen S K, Boschung J, Nauels A, Xia Y, Bex V and Midgley P M. Cambridge University Press, Cambridge, United Kingdom and New York, NY, USA.

Paula M A, Reis V M and Döbereiner J (1991) Interactions of *Glomus clarum* with *Acetobacter diazotrophicus* in infection of sweet potato (*Ipomoea batatas*), sugarcane (*Saccharum* spp.), and sweet sorghum (*Sorghum vulgare*). *Biol. Fert. Soils* 11, 111-115.

Pirttilä A M, Laukkanen H, Pospiech H, Myllyla R, and Hohtola A (2000) Detection of intracellular bacteria in the buds of Scotch pine (*Pinus sylvestris*) by *in situ* hybridization. *Appl. Environ. Microbiol.* 66, 3073-3077.

Reay D S, Davidson E A, Smith K A, Melillo J M, Dentener F and Crutzen P J (2102) Global agriculture and nitrous oxide emissions. *Nat. Climate Change* 2, 410-416.

Richards M and Sander B O (2014) Alternate Wetting and Drying in Irrigated Rice: Implementation Guide for Policymakers and Investors as Practice Brief on Climate Smart Agriculture (CGIAR Research Program on Climate Change, Agriculture and Food Security, Copenhagen).

Rocafull Y R, Badia M J, Garciea M O, Álvarez B D and Sánchez J R (2016) Isolation and characterisation of *Gucoancetobacter diazotrophicus* strains. *Cultivos Tropicales* 37 (1), 34-39.

Rogers C, and Oldroyd G E D (2014) Synthetic biology approaches to engineering the nitrogen symbiosis in cereals. *J. Exp. Bot.* 65 (8), 1939-1946.

Schuller G, Hertel C and Hammes W P (2000) *Gluconacetobacter entanni* sp. nov. isolated from submerged high acid industrial vinegar fermentations. *Int. J. System. Evolution. Microbiol.* 50, 2013-20.

Sevilla M, Burris RH, Gunapala N and Kennedy C (2001) Comparison of benefit to sugarcane plant growth and ^{15}N corporation following inoculation of sterile plants with *Acetobacter diazotrophicus* wild-type and *Nif*-mutant strains. *Mol. Plant Microbe Interact.*14 (3), 358–66.

Shewry P R (2007) Improving the protein content and composition of cereal grain. *J. Cereal Sci.* 46, 239–250.

Simmonds N W (1995) The relation between yield and protein in cereal grain. *J. Sci. Food Agric.* 67, 309–315.

Simmonds J (2008) Community Matters: A history of biological nitrogen fixation and nodulation research 1965-1995. PhD Thesis Rensselaer Polytechnic Institute, Troy, New York. UMI Number 3299478.

Smil V (1999) Denonator of the Population Explosion. *Nature* 400, 415.

De Souza L E, De Souza S A, Oliveira M and Ferraz T M (2015) Endophytic colonisation of *Arabidopsis thaliana* by *Gluconacetobacter diazotrophicus* and its effect on plant growth promotion, plant physiology, and activation of plant defence. *Plant Soil* 399 (1).

Snyder C S, Bruulsema T W, Jensen T L and Fixen P E (2009) Review of greenhouse gas emissions from crop production systems and fertiliser management effects. *Agr. Ecosyst. Environ.* 133, 247–266.

Stasiak L and Błażejak S (2009) Acetic acid bacteria - perspectives of application in biotechnology - a review. *J. Food and Nutr. Sci.* 59 (1), 17-23.

Stephan M P, Oliveira M, Teixeira K R S, Martinez G and Döbereiner J (1991) Physiology and dinitrogen fixation of *Acetobacter diazotrophicus*. *FEMS Microbiol. Letts.* 77, 67-72.

Sutton M A, Howard C M, Erisman J W, Billen G, Bleeker A, Grennfelt P, van Grinsven H and Grizzetti B (Eds) (2011) The European Nitrogen Assessment. Cambridge University Press 664.

Thomas P and Reddy M K (2013) Microscopic elucidation of abundant endophytic bacteria colonising the cell wall - plasma membrane peri-space in the shoot-tip tissue of banana. *AoB PLANTS* 5, plt011; doi:10.1093/aobpla/plt011.

Thomas P and Sekhar A C (2014) Live cell imaging reveals extensive intracellular cytoplasmic colonisation of banana by normally non-cultivable endophytic bacteria. *AoB PLANTS* 6, plu002. doi.org/10.1093/aobpla/plu002.

Tilman D, Cassman G K, Matson P A, Naylor R and Polasky S (2002) Agricultural sustainability and intensive production practices. *Nature* 418, 671–677.

Toft C and Andersen G E (2010) Evolutionary microbial genomics: insights to bacterial host adaptation. *Nat. Rev. Genet.*11, 465-465. DOI: :10.1038/nrg2798.

Tucker M (2004) Primary nutrients and plant growth. In: SCRIBD, editor. Essential plant nutrients. NC, USA. North Carolina Department of Agriculture; 2004, 1–9.

UNEP (2007) UNEP and WHRC *Reactive Nitrogen in the Environment: Too Much or Too Little of a Good Thing* (United Nations Environment Programme, 2007).

White J F, Torres M S, Johnson H, Irizarry I, Chen Q, Zhang N, Walsh E, Somu M P, Bergen M (2013) Intracellular colonisation and oxidative lysis of bacteria in vascular plant seedling tissues. July 2013. DOI:10.13140/RG.2.1.3732.6880.

Wu L L, Yuan S and Huang L Y (2016) Physiological mechanisms underlying the high-grain yield and high-nitrogen use efficiency of elite rice varieties under a low rate of nitrogen application in China. *Front. Plant Sci.* 7,1024. DOI: 10.3389/fpls.2016.01024.

Zhang, S (2016) A Chemical reaction revolutionised farming 100 years ago. Now it needs to go. SCIENCE June 2016.

Rice results

Azotic have published a number of on-line news items and pdf documents which it has used for promotional purposes:

Azotic Technologies 2018 Azotic's Natural Nitrogen Fixing Technology is now commercially available in the USA. 10/12/2018

Azotic Technologies get positive results from rice trials. 17/09/2018

Azotic Technologies Update Brochure.pdf. 17/09/2018

Natural Nitrogen Fixation for Sustainable Agriculture.pdf. 20/06/2018

www.ingramcontent.com/pod-product-compliance
Lightning Source LLC
Chambersburg PA
CBHW070544090426
42735CB00013B/3069
* 9 7 8 1 9 1 6 2 1 8 1 0 9 *